Territory, Globalization and International Relations

Territory, Globalization and International Relations

The Cartographic Reality of Space

Jeppe Strandsbjerg
Copenhagen Business School, Denmark

First published 2010 by
PALGRAVE MACMILLAN

Palgrave Macmillan in the UK is an imprint of Macmillan Publishers Limited, registered in England, company number 785998, of Houndmills, Basingstoke, Hampshire RG21 6XS.

Palgrave Macmillan in the US is a division of St Martin's Press LLC, 175 Fifth Avenue, New York, NY 10010.

Palgrave Macmillan is the global academic imprint of the above companies and has companies and representatives throughout the world.

Palgrave® and Macmillan® are registered trademarks in the United States, the United Kingdom, Europe and other countries.

ISBN 978–0–230–58083–1   hardback

This book is printed on paper suitable for recycling and made from fully managed and sustained forest sources. Logging, pulping and manufacturing processes are expected to conform to the environmental regulations of the country of origin.

A catalogue record for this book is available from the British Library.

Library of Congress Cataloging-in-Publication Data

Strandsbjerg, Jeppe, 1973–
     Territory, globalization and international relations: the cartographic reality of space / Jeppe Strandsbjerg.
          p. cm.
          ISBN 978–0–230–58083–1 (hardback)
     1. Human territoriality—Political aspects.   2. Geopolitics.
3. International relations.   I. Title.

     JC319.S77 2010
     327.101—dc22

                                                        2010027563

Printed and bound by CPI Group (UK) Ltd, Croydon, CR0 4YY

To Kjeld Anker Kabel (8.11.1914–8.5.2009);
a humorous humanist with a curious interest in anything
worldly and nothing spiritual.

Contents

List of Maps

Acknowledgements

I first got the idea that came to frame this book in a discussion about state identity and 'the international' on a sunny day drinking coffee under the big cherry tree outside Arts D at Sussex University. From that moment until now, many people have made their mark on the argument either through engaging criticism and discussion, in the role of discussants at conferences and seminars; by critically responding to drafts; or a combination of the three. I would like to thank, gratefully, the following people for their contribution to retrieving the idea of this book from *terrae incognitae* and bringing it back to the familiar landscape of academic writing: Jens Bartelson, Gurminder K. Bhambra, Michael Biggs, Martin Coward, Henrik Dupont, Matthew Edney, Stuart Elden, Huw 'Flairs' Evans, Janus Hansen, Clemens Hoffmann, Beate Jahn, Lars Bo Kaspersen, Jochen Kleinschmidt, Sam Knafo, Rhiannon Lambert, Henrik Gutzon Larsen, Tobias Lindeberg, Kasper Lindskow, Kamran Matin, Christiane Mossin, Arthur Mühlen-Schulte, Alexander B. Murphy, Daniel H. Nexon, Mia Nyegaard, Mihnea Panu, Raluca Parvu, Fabio Petito, Reuben S. Rose-Redwood, Justin Rosenberg, Len Seabrooke, Ole Jacob Sending, Martin Shaw, Robbie Shilliam, Grahame Thompson, Antje Vetterlein, Duncan Wigan and Uffe Østergård. The anonymous reviewer provided excellent feedback and suggestions for improving the book. Some of the material in Chapter 6 has previously been published in Strandsbjerg, J. (2008). 'The cartographic production of territorial space: mapping and state formation in early modern Denmark'. *Geopolitics* 13(2): 335–58, and an earlier draft of Chapter 5 has been published as Strandsbjerg, J. (2008). The cartographic assemblage of the globe. Working paper no 45. International Center for Business and Politics, Copenhagen Business School. I would also like express my gratitude to the Royal Library, Copenhagen, which has generously allowed me to reproduce their digitized historical maps. The Centre for Business and Politics at the Copenhagen Business School has provided a fertile academic environment (especially our many seminars where my brilliant colleagues have patiently commented on my cartographic stories), and the Geography Department at Durham University provided a kind refuge allowing me to finish this manuscript. At Palgrave I am grateful for the patience and helpfulness of Philippa Grand, Olivia Middleton, and Manavalan Bhuvanaraj. The Danish Research Agency generously funded some of the research that has gone into this book. Finally to Sara: thanks for love, patience, disciplining inquiries and, not least, being there. Question is whether we will ever decide what kind of space 'there' is .... And Elmer, well, you have probably been a bigger obstacle than support; not sure that academics are really supposed to have children. But you are there to remind me of other things everyday; not least balloons and caterpillars!

1
Introduction

Under the heading 'On the Threshold of a New Era', we read how 'the breathtaking speed of contemporary technological progress has revolutionized our concepts of time and distance' (Krarup Nielsen et al. 1930: 9, my translation). Titled 'The Conquest of the Earth' (*Jordens Erobring*), six large volumes tell the history of how 'the white race' has strived to expand its knowledge and rule over foreign lands and people – through trade, exploration, conquest or scientific expeditions (Krarup Nielsen et al. 1930: 15). The authors follow the line set by the notorious geopolitician Halford John Right Mackinder when they note how the last 50 years of polar expeditions (at the time of writing) represent an effort to complete the conquest, or closure, of the world (Krarup Nielsen et al. 1930: 23). As techno-mechanical innovations penetrate the most remote sites of the earth, customs and tradition are demolished as we strive towards human uniformity.

These propositions were formulated in 1930, yet they are strikingly similar to current arguments about globalization. They illustrate a sense of technological awe, globalization and a concern with identity politics which are also central to current debates. It further represents a concern with eras, and the idea that the current state of affairs somehow represents the threshold of a new era that is challenging established knowledge of the political world. Today we are supposedly living in a globalized, post-cold war world – and maybe even in one that is postmodern. Since the end of the cold war, codified by the year 1989, there has been an increasing preoccupation with change and attempts to grapple with an organization of the world seemingly very different from that of the not so distant past. Global space is allegedly being reconfigured; previously stable identities are being rendered uncertain and the way in which cultural difference is being negotiated is taken out of the supposedly concrete foundations of the nation state and thrown into a maelstrom of issues related to migration, networks, mobility and increasing temporality. The notion that technological progress alters the role, and often the significance, of geographical space has stimulated a debate concerning the spatiality of contemporary politics. As social relations, allegedly, are no longer contained by the territorial confines of the nation state, one gets the impression, quoting Marx and Engels, that 'all that is solid melts into air' (Marx and Engels 1998: 6).

Key to the globalization literature is an understanding that social practice is moved away from the spatiality of the state towards a larger global setting, and sometimes, concurrently, towards the smaller regional scale. At the same time, a new kind of postmodern subjectivity characterized by mobility has been articulated as being typical of our time. Generally, the unsettling of space in relation to the particular field of investigation has called for a reconsideration of culture (Gupta and Ferguson 1992), society (Urry 2000) and the state (Walker 1993), as well as, our historical understanding of the development of politics, imperialism and colonialism. As an example, John Urry has promoted the idea of a shift in sociology from seeing the 'social as society' to seeing the 'social as mobility' as national borders are being permeated by growing networks and flows (2000: 1–3; 2007). The social as mobility entails a different conceptualization of social relations, and it denotes a different spatiality than that associated with national societies. In a similar vein, Nigel Thrift (1996) has argued how new (compared to known structures of the nation state) spatial formations emerge as a result of mobility becoming the central feature of social life. In the literature speaking directly to the conceptualization of globalization, Jan Aart Scholte suggests that globalization entails that a supraterritorial dimension has developed alongside a traditional territorial social geography (2005: 60–77).

As the literature on globalization is mushrooming, there are, of course, a myriad conceptualizations (for a recent taxonomy, see Hirst et al. 2009), though, the main aspect of this phenomenon that this book will address is the notion of spatial change and the degree to which a non-territorial logic overtakes the territorial order associated with nation states. In short, this can be described as a tension between networks and territories. As such, the notion of globalization has led to two sets of questions reverberating through the social sciences at large. The first set of questions address the spatial particularity of the modern state: why and how has the territorial state become the dominant political form; how are societies territorialized; and does this occupation prevent us from seeing non-territorial socio-political forms taking shape because we are blinded by established categories? The second set of questions concern the understanding of space, and on a very general level, the relationship between social practice and space: how did we come to understand space in absolute terms; how can space be reconceptualized more adequately; and how is space produced in various historical settings? Whereas the former set of questions has led to concern with state formation in a historical perspective (we must know where we come from to understand the possibilities of future transition), the second set of questions is tied up with an interest in different conceptions of space and different social spatialities. These are, then, the questions that need to be answered in order to discuss changes to the political organization of space.

In Edward Soja's original formulation, 'the political organization of space' refers to the way space, and human interaction in space, are structured to fulfil political functions (1971: 1). In a similar vein, I understand the political organization of space as processes of establishing structures of governance within particular spaces. Both the ability to demarcate space as well as constituting 'the world' within which the individual space is demarcated are vital to organize and govern space. Based on this conceptualization, the purpose of this book is to investigate how the questions raised above have been articulated in the social science literature that poses globalization as a challenge to the state. This is a debate which has been prevalent in sociology and geography but the question has posed a particular challenge to the discipline of International Relations (IR), which will therefore provide the main disciplinary focus of the book. Traditionally IR has been a discipline studying relations between state actors; even if this picture has been challenged by institutions mediating the interests and agency of states, or by ideas of World Systems of production from which the state system is derived, it is the multitude of states which has provided the *raison d'être* of the discipline. In a sense, sociology would still be sociology, and human geography would still be human geography if social relations were to be reconstituted beyond the confines of territorial states. But if the state ceased to play any meaningful role in social life, what then would IR be?

Few suggest today that the state will wither away in the face of globalization, but due to its centrality the question of the political reorganization of space provides an important theoretical challenge across the social sciences, in general, and to IR, in particular. This becomes clear when reading through the literature that picks up the challenge. As I will demonstrate in this book, most accounts are lacking in bringing together and unfolding these questions in a manner that can adequately grasp fundamental changes to social spatiality as a consequence of globalization.

The key problem across the writings concerned with a changing political organization of space is exactly the failure to historicize space itself. In most attempts to do this, I will argue, space is either fitted into an existing narrative of historical sociology broadly understood, or it is reduced to a textual phenomenon that can be adequately grasped through discourse analysis and meaning creation. As a result the literature tends to maintain a problematic historical understanding of a transformation from a medieval order to a modern territorial order characterized by different spatialities, while, at the same time, implicitly maintaining a profound subject/object distinction with regard to the conceptualization of space. That is, space is either rendered social or natural. I suggest that to tackle the issue of the political reorganization of space, it is necessary to historicize space and the state at the same time. Hence, maintain them both as historically contingent phenomena. In effect, this argument is a call for a more complex understanding of the historicity and nature of political space in the assessment of change associated with globalization. In the attempt to overcome simple historical oppositions concerning a transition from medieval to modern state-spaces now withering away in the face of global networks and flows, I employ what can be seen as a historical sociology of space formation that takes seriously the interrelationship with the development of political organization or, put simply, state formation. In the attempt to do that, the main spatial theme at stake is the relationship between territory and globalization reverberating around a discussion about what international relations is, and how it has developed as a practice between spatially defined subjects operating on a global stage.

The main argument driving this book is that we live in a cartographic reality of space. This is to say that the spatial reality underwriting state territory, globalization and the conduct of international relations is assembled through cartographic practices historically and we need to understand this before we engage in discussions concerning spatial change of the state. I argue that the so-called cartographic (and scientific) revolution that took off in the European renaissance initiated a process through which space was assembled as a unitary, yet abstracted, reality that served as the basis of state formation as well as the establishment of colonial empires and trade networks. In effect, it was a transition in cartographic practice in Europe that altered the reality of space and hereby altered the conditions for producing political space. Following from this, I argue that territorial and global space rely on the same cartographic reality and, therefore, it is somewhat flawed to posit state territory and the global as *spatial* opposites and globalization necessitating a reformulation of the role of space in social theory.

Empirically, the argument is supported by an analysis of cartographic practice in the early modern period (ca. 1450–1650) which is said to represent a transformation in the concepts of space and time as well as the political organization of the world. This period was witness to a dramatic reconfiguration of space on a global scale, both in terms of social power and knowledge. Politically, European powers emerged as dominant in the world, and ideas of sovereignty, the decline of universal authority, an emerging understanding of national identity and so forth all used, and contributed to developments in, cartography. At the same time, previously held dogmas and assumptions that underpinned the late medieval world view were challenged. Much has been written about how the 'discovery' of a new continent in 1492, America, posed a challenge to the authority of the ancient sources and the Bible as the foundations for knowledge. Not least the universality of knowledge and authority was questioned, and identity and cultural difference took on a new meaning (Jahn 2000; Pagden 1993; Mignolo 1995: 317). Such a historical perspective will provide a basis for comparing current claims to changes of the political organization of global space.

More specifically, the analysis concentrates, first, on the production of global space as a coherent political space, and second, on the formation of Denmark as an absolutist state. In the analysis, cartography is not ascribed causal explanatory value, but rather, the analysis will show how the establishment of cartographic spaces preceded and enabled a novel relationship between space and politics. Where England, France, Spain and the Habsburg possession have been thoroughly scrutinized, Denmark is often overlooked in the literature on state formation, which is a shame for two reasons. By 1660, in the words of Thomas Ertman, Denmark turned into 'a paradigmatic case of absolutism' (1997: 310), which, according to the canon of IR, is considered the beginning of modern international relations. And even if this has been rejected, it has to a large extent been the ideal of a coherent territory with a single centre of authority which has pervaded the IR notion of territory (Teschke 2003: 1). Yet, more importantly, Danish state formation in the said period represents an interesting case of the integration of a relatively self-contained Baltic political system into a European stage of interstate relations. At the beginning of our period, various polity types were competing, while in the mid-seventeenth century the territorial state had come out 'on top' at the expense of the Hanseatic league, a Nordic Union and self-governing areas, such as the area of Ditmarsken. Located at the margins of the Holy Roman Empire, Denmark interacted with, but was also somewhat autonomous from, the Emperor, and this transition to a single type of political organization, the sovereign territorial state, represents a significant dimension of the history of state formation especially with regard to the IR discipline.

The spatial identity of the state

If globalization is challenging the traditional sovereign state as the predominant way of organizing space politically, then this calls into question how the state was organized as a spatial entity in the first place. In other words, this is a historical question about the spatial identity of the state and how the state came to be understood as a spatial thing abstracted from the immediate social relations and networks of the everyday running of state business. Identity, in this context, should not be understood in terms of national identity and the degree to which a population identifies with a territorially defined polity, but rather on an abstract level concerning what spatially constitutes the state as a state. This points to the role and status of 'territory'. Posing this as a question of spatial identity indicates that there is more at stake than simply establishing state institutions against, or within, an existing territory. In the literature on political geography it appears that territory and territorial behaviour has been an important aspect of social practices in much of human history (Gottmann 1973; Sack 1986; Soja 1971; Delaney 2005). This somewhat contradicts the widespread notion in IR theory that territorial politics is a modern phenomenon codified by the peace treaties of Westphalia in 1648 and indeed when browsing the historical catalogue of wars – colonial, European and others – it seems that they were often about the control over space, territory, domain, demesne, dominion, rights of passage, rights of use, rights of taxation and so forth. This suggests that territorial issues, and thus the question of territory, are always relevant when analysing polities and their interaction. What is important for the understanding of international relations and the so-called modern state is the question of how it became possible to think and practice sovereignty in spatial terms. That is, how was it possible to relocate sovereignty from the ruler of a state to a fixed notion of the territory?

This question has often been discussed as a question of developing clearly defined borders but, as I will discuss subsequently, the focus on borders as that which constitute political spaces ignores how different ways of establishing space decides how borders can be invoked. This implies that something happened to space, historically, that made it possible to define and demarcate sovereignty in spatial terms. While sovereignty has always had a relation to the demarcation of jurisdiction and, thus, contained a reference to space, it was not always the relationship between sovereignty and space that decided the boundaries of social relations. The key feature to understand is the transition from what Soja has called a social definition of territory to a territorial definition of society (1971: 13). In the Middle Ages in Europe, for example, sovereignty would be secured through social bonds of allegiance, and it was the possessions of the king as well as those considered loyal to the king that would define the scope of the territory. In this situation, it is the social relations of allegiance that define the territory and not the other way around.

To capture this transition ought to be a central concern for social science in general, and IR in particular; nevertheless there have been surprisingly few attempts to deal with the issue of territory as a spatial problem, by which I mean, that the spatiality of territory is often assumed rather than investigated. Generally speaking, territorial space is more often than not either taken as a given physical backdrop to the state, or rendered an inadequate product of modernist knowledge by social constructivist critics. This is somewhat peculiar, in that critical theory, postmodernism and feminism have pushed 'the field to a greater philosophical and political self-consciousness' (Inayatullah and Blaney 2004: vii). Even though this has materialized in significant philosophy of science debates as disputes over the character of political communities, this push towards further reflection on the core concepts of the discipline has to a large extent left behind the concept of space. While Inayatullah's comment is aimed at an IR audience, it neatly captures how the self-reflexivity and philosophical debates concerning the status of 'science' have left space caught in between physical reality and an abstract, often textual, world of social construction.

For those believing in an objective reality, space is seen as nature and mapping space is just a matter of getting to know this reality. In this understanding, territory is a physical background which, in its quality of being nature, is deemed to lie outside the realm of social scientific conduct. On the other side of this divide, approaches informed by social constructivism, in one guise or another, tend to treat territory as something that is generated through social discourse, and that is often, textual constructions. By showing how territory is a result of social and historical meaning constructions, social constructivism entails a notion of how things could be different. However, because space is rendered a social construction it is very hard to see what limits these possible new constructions. In writing any materiality out of social constructions of space, these approaches, in a sense, take the spatial element out of territory, and in effect, reinstate a solid divide between territorial space seen as natural geography (realists) or as culture (social constructivists). As such the divide between realism and social constructivism seems to follow and reproduce the nature/culture divide. The problem with this is that any notion of territory without a spatial referent is never going to be too convincing. Regardless of how it is formulated, territory entails a relationship between a human collective and the environment. If space does not have any meaning in and of itself, a physical notion of space considered to lie outside the social domain is no good. Neither will it suffice to discuss space purely in terms of 'conceptions', because as Lefebvre (1991) reminds us, the conception of space is only part of its reality.

The solution I propose is to be more constructivist than social constructivism usually is. In that, the theoretical ambition of this book is to posit cartography as a social practice that exactly serves to establish space as part of a 'social reality' – for lack of a better term – yet, at the same time, as a social practice which cannot be understood without including the relationship between people and the environment. In that respect, the study of cartography provides a possibility for studying the relationship between culture and nature, between what is usually deemed subjective and objective. In effect, the study of cartography both constrains and expands constructivism. It constrains constructivism by reintroducing a notion of reality, which is real by virtue of being constructed, and it expands constructivism by taking it into the realm of nature. These arguments leave three issues to be resolved before we can proceed. First the question of the relationship between space and state formation; second, how cartography can be seen as 'space making'; and finally, the theoretical implications of seeing space as being constituted through cartographic practice.

State formation and cartographic space

The first question concerning the relationship between space and the state has been explored in the junction between human geography and historical sociology, and there has been extensive writing on state formation in general terms. Yet for the sake of this introduction it is important to show how space is missing in the established accounts of state formation because this illustrates the problem concerning territorial space and why space has yet to be included in the analysis as something whose transformation itself was an important aspect of state formation.

By historicizing political structures, Historical Sociology has problematised, conceptually elucidated and historically empirically investigated the sovereign territorial state which is conventionally taken for granted as the starting point for IR (see Lawson (2006) for a comprehensive discussion). However, as Jens Bartelson has argued, macrosociology struggles to unveil adequately the historical politicization of space and the spatialization of politics (1995: 40). This is odd as the notion of territory plays a pivotal role for the understanding of the state. In Max Weber's own words: 'a state is that human community which (successfully) lays claim to the *monopoly of legitimate physical violence* within a certain territory, this "territory" being another of the defining characteristics of the state' (1994: 310–11). This emphasis on centralized authority within a given territory is a central aspect of the Weberian strand within historical sociology.

In Charles Tilly's celebrated study of state formation in Europe, he defines states as 'coercion-wielding organizations that are distinct from households and kinship groups and exercise clear priority in some respects over all other organizations within substantial territories' (1992: 1). Territory is a significant defining feature of the state, and also, part of the struggle out of which states emerge: 'states always grow out of competition for control over territory and population', and hence they invariably appear in clusters, and form systems (Tilly 1992: 4). Tilly describes the process by which state rulers established coercive means, and deprived the civil population of the same, fortified boundaries and thus made clearer divisions between internal and external politics, and he suggests that Weber's notion of the state started to make sense up through the seventeenth century; or that we start to recognize this image of the state (1992: 68–70). However, the processes Tilly describes are taking place on a smooth surface of space – a surface which can be bounded and controlled, but he remains silent on the processes of 'planning this surface'. Territory is considered a generic thing containing the same meaning over time.

We see a similar trend in the magnum opus of Michael Mann, *The Sources of Social Power* (1986; 1993). Although Mann conceptualizes society as a socio spatial network of power (1986: 1) and engages with the spatial issues of the state, he assumes that the state is territorial without developing a discussion about what kind of space territory is. Space is something in which social action takes place and something that politics try to dominate. Political power is necessarily centralized and territorial, and revolves around a dual spatiality: domestically, political organization is territorially centralized and bounded, whereas externally, political organization is geopolitically organized in a system dominated by such features as empire, or a multistate civilization (Mann 1986: 11, 27).

Mann provisionally defines the state (modifying Weber) as 'a differentiated set of institutions and personnel embodying centrality, in the sense that political relations radiate outward to cover a territorially demarcated area, over which it claims a monopoly of binding and permanent rule making, backed up by physical violence' (1986: 37, italics removed). He rightly emphasizes that there is no 'single bounded society in geographical or social space' (Mann 1986: 1), and that different state forms require different territorialities. For example, Mann argues that the *ständestaat* requires a more 'solid' and universal territorial space compared to the feudal state because it depends on the territorial coordination of autonomous actors (1986: 445). As also pointed out by Felix Driver, there are clear spatial connotations in Mann's concept of infrastructural power (1991: 273). Nevertheless the difference between state forms is the degree of territorial coordination, and the clarity of boundaries; the *ständestaat* requires more solid boundaries than the feudal state and so on. Thus, he abstains from a discussion about how political power interrelates with space. What remains is this predominant notion that space is a stable playing field for social practice.

Finally, Anthony Giddens also works with a Weberian definition of the state, and possibly takes it a step further than Tilly and Mann in historicizing the territoriality of the state because he identifies the changing notion of political boundaries as being a crucial feature of the transition from traditional (or feudal) states to the modern state. He defines the state as 'a political organization whose rule is territorially ordered and which is able to mobilize the means of violence to sustain that rule' (Giddens 1985: 20). Crucial to Giddens' argument is the emphasis on the transition from frontiers to boundaries for the development of the modern state. The territorial form of the state is accentuated through the internal consolidation of the state, 'and it is during the period of absolutism that Europe became altered in respect of states' boundaries' (Giddens 1985: 85). According to Giddens, territories of medieval rulers were not necessarily continuous but scattered: '[t]he centralization of political power associated with absolutism was not a simple process of the expansion of effective control over areas already nominally subject to the authority of the ruler. It demanded substantial alteration in the external and internal frontiers of states' (Giddens 1985: 89). The impression left is that territorial regulation, as the key to political power, is about constructing boundaries in space considered as a 'flat surface'. Despite the discussion of spatial relations, there is a profound lack of attention given to how space itself is transformed over time, and how this spatial transformation played a significant role in state formation.

This critique of the state formation literature parallels Stuart Elden's suggestion that territory 'is unhistorically accepted, conceptually assumed and philosophically unexamined' (2005a: 10); and that the crucial feature to grasp about the notion of modern territory is its condition of possibility. He argues 'that it is the understanding of political space that is fundamental and the idea of boundaries a secondary aspect, dependent on the first' (Elden 2005a: 11). Analysing modern state formation, the primary focus ought to be on space, rather than on boundaries, and it is therefore necessary that we seek to historicize knowledge of space because space is not a stable category. In consequence, to understand the modern notion of territory the construction of a specific notion of space should have primacy relative to the idea of boundaries. My argument here is that 'a territorially demarcated area' is not simply about demarcation and boundaries, but about the production of space itself; how it was possible to draw boundaries in a specific way; and what kind of space was 'contained' within these boundaries.

This implies that the transformation from one state form to another, or from one societal form to another, is not simply a matter of degree of territorial control. It is about territorial space itself and how its meaning is produced. Territory and transformation are therefore – and significantly also – a question of how the space is produced and recognized as space in and of itself. Thus, it is not only in the 'shape of territory' that space is political. As Bartelson argues, space 'turns political' in its recognition as a fact (1995: 31). With a few exceptions (for example Harste 2000), Weberian historical sociology remains silent about the spatial transformations which accompany the development of the sovereign territorial state. The point here is not to call for a replacement of the existing narratives on state formation, but rather to point out an aspect which is generally missing from these explanations. And this is where cartography makes it possible to give a measure of analytical autonomy to the production of space without rendering space a static or essential dimension.

Moving on to the second question, outlines above, it is possibly because maps are a 'given' in most people's everyday life, that mapping seems to be a natural and uncontroversial exercise. Maps and the cartographic image of the world are so common that they are, in effect, second nature. Very few think that maps could look differently, and that the cartographic image of the world is neither a necessity nor natural. In an attempt to unsettle this taken-for-granted status of map-making, recent studies have investigated the power of maps and mapping. Foucault-inspired analyses have highlighted the power of maps by pointing out that even scientific maps represent a reality that we 'think we know' and not some essential, or given, reality (e.g. Wood 1992), and the field of historical cartography has been reinvigorated as a social studies discipline through the pioneering work of Brian Harley and David Woodward (Harley and Laxton 2001; and the impressive History of Cartography project, Harley 1987; Woodward 2007b). Writings by John Pickles (2004), Michael Biggs (1999) and Mark Neocleous (2003) have all emphasized the significance of cartography for what could be labelled the modern political organization of space. They share a notion that cartography plays a constituting role for spatial conceptions and practices, and thus also, state formation. And this is, to a degree, to ascribe some measure of agency or causality to cartography with regard to the relationship between social practice and space – even if it is not articulated in these terms. In a similar manner, the argument of this book draws extensively on what could be labelled a 'critical history of cartography'.

More specifically, the theoretical framework for understanding maps, and their history, builds primarily on the Foucault-inspired writings of Brian Harley (Harley and Laxton 2001; Edney 2005). I suggest that we should be aware of two dimensions concerning mapping: discursive power and the power of authorship. The one concerns epistemic principles informing map-making, which in effect decide what is considered a map, and this decides how space is known. The other concerns the agency of map-making; who draws them, and what is the institutional organization behind it. I combine this reading of cartography with a broader discussion concerning the social nature of space in order to establish the argument that we live in a cartographic reality of space.

The argument here is not one that focuses on propaganda maps and those maps very explicitly misrepresenting 'reality' for a specific political purpose. The argument unfolds on a more abstract level and is concerned with a general relationship between space and cartographic practices. The study of cartographic practice enables a study of how space is conceptualized in different places and at different times. The period from 1450 to 1650 was witness to what has been dubbed a 'cartographic revolution' in Europe changing the ways in which space was represented and conceived. The analysis will focus on how this cartographic transition played out in the European context where it became an essential part of European expansion and conquest of the rest of the world. I will seek to demonstrate how this transition facilitated new ways of controlling territory, but also how the epistemic power of the European map produced space in a way that played a significant role for developing forms of political organization in Europe. The argument is that the new mode of modern, or more precisely, geometric cartography[1] altered the reality of space, and hereby altered the conditions and possibilities for political organization.

Finally, on the third question concerning the theoretical implications, the focus on cartography provides a way of studying the production of space without making space either a physical constant or something that is derived from general narratives such as 'capitalism' or 'war'. That is, to make cartography a somewhat autonomous object of study. While the danger of this is to be blind to the extent that a historical transition of cartographic theory and practice was part of a wider social transformation, the advantage clearly is to save space formation from metanarratives about social development, and this enables a specific focus on practices of concrete space-making. And this facilitates an analysis of the relationship between space and political change without necessarily making historic space-formation a product of the social processes it is analysed in conjunction with, as for example, state formation. In other words, to study space through cartography makes it possible to maintain space as an autonomous phenomenon in the analysis. To illustrate the necessity of this, globalization theory is a useful venue due to the profound link between the intellectual problem of globalization and the imagination of time and space as empty, homogeneous and abstract (Rosenberg 2000: 5). The link between the abstraction of time and space from how they are subjectively experienced is a theme running throughout discussions of modernity and a possible postmodern era of globalization (Giddens 1979, 1990; Sack 1980; Soja 1971; Kern 1983). The modern territorial state can be said to rely on an abstraction of space to the extent that the notion of sovereignty indicates a relationship between a territorial space and a centralized government whose claim to legitimate authority is demarcated in spatial terms. And in that respect, political space comes to be abstracted from particular, or local, social experiences and reconfigured in terms of a universal notion of particular sovereign space. This abstraction has furthermore erased differences within the territory and constructed territory as a universal homogeneous space within its boundaries. Such abstraction and reconfiguration is obviously evident in the drawing of those notorious colonial borders in Africa which follow the path of a ruler rather than any social formations which might have existed prior to the drawing of these boundaries.

This link between a specific conception of space as homogeneous, continuous and empty with the modern development of capitalism and the state is well established (Harvey 1989; Ruggie 1993; Walker 1993). However this literature, as well as the literature on historical sociology, rarely examines how the abstraction of space has taken place in practice.

This is important because it is not a sufficient explanation to observe that an abstract conception of space materialized alongside these processes of state formation. As Bruno Latour has argued, social action cannot take place in an abstract space (1999: 28), and it is, then, exactly the specific production of space as an abstract, yet real, category that should be the object of investigation. This is a crucial point because the organization of global capital, and the international system of sovereign states may all rely on, and support, a notion of space abstracted from everyday perceptions, but this is not a sufficient explanation of how space itself has been rendered abstract.

If action cannot take place in abstract space, space has to be made abstract before the political economy of territorial states, and capitalism can be organized and consolidated. Action cannot be planned and coordinated in an unknown space: it is not possible to organize a meeting in any location without a shared intersubjective understanding of this place, or knowledge about how to get there. Hence, knowledge of space is crucial not only for how we think about space but also for how it is possible to organize it. Spatial knowledge is not simply a matter of conceptions of space, but also of agency which point to how the physical landscape is integrated into a system of thought and a social order. Cartography has done exactly that; rather than being a curious and innocent exercise helping us to navigate (though it also does that), cartography establishes a spatial reality which, in turn, conditions possible social action. In the context of the political and economic history of early modern Europe, I will argue, it is a specific mode of cartographic practice which turned space into a well-ordered abstraction that enabled the planning of socio-economic action as if space was indeed homogeneous, continuous and empty.

This argument poses another question concerning the role of cartography in relation to the broader transformation occurring in early modern Europe. The transition in the theory and practice of cartography was part of a wider socio-political transition. Several authors, such as John Hale, point to a new way of seeing and representing space in visual arts as an expression of a new approach by Europeans to the world more generally (1971: 51). The emergence of linear perspective in this period entails a dramatic change in the representation of space as regards, for example, visual arts, fortification and cartography (Blunt 1968; Edgerton 1975; Jammer 1969). It is thus important to stress that the significance of cartography is not isolated from more general transformations, and thus, when I suggest that a cartographic transition conditioned and made possible the particular organization of the world that we know today, the question inevitably begs itself: what made cartography possible in the first place? If cartography was actually made possible by broader socio-economic developments, one could argue that the notion that it was cartography and not war or capitalism that produced space in such a way that it became possible to create a system of sovereign territorial states would have to be significantly modified.

Here, however, it is important to remember two things. First, I do not to ascribe any causality to the cartographic developments in terms of causing broader social change in itself. Second, I do not believe that it will tell us much to posit cartography simply as an outcome of wider social transformation. Of course geometric cartography did not invent itself, but it was introduced in a setting where it entered into a relationship with socio-political changes and came to play a significant role in those changes. The demise of the universal powers of pope and emperor, the needs of exploration and navigation and the ascendance of particular states within Christendom all created a demand for the new cartography. However, these changes do not in themselves explain the emergence of a new way of representing space. As stressed by Hale, the domination by Europeans was made possible in particular by a development in geographical theory and a shift in the way that 'men' imagined terrestrial space (1971: 47–9) but this is not the same as saying that these political developments caused the development of a new mode of cartographic practice.

In sum, to maintain an autonomous view on cartography allows me to investigate both how novel cartographic principles established space in a novel fashion and, also, to trace the agency in terms of how cartography was adopted by emerging empires and territorial states respectively. That will provide a more nuanced picture of the role of cartography than simply being an immanent feature of broader socio-political change. It was cartography specifically that established and transformed the material environment in a way that made overseas planning and coordination of space possible in a way it had not been before. It was cartography that rendered space autonomous and made possible a territorial definition of society that had not existed before. As such, it was not technologies of communication, production or warfare that provided the spatial underpinnings of a territorially organized state system within a unified global space, as it is often claimed, but rather a specific knowledge technology of space. Cartography did not invent territory but it changed the ways of producing territory, and in that respect cartography conditioned and facilitated the formation of states, empires and a global space during the early modern era. Although it was not cartography that initiated or completed these changes, it is impossible to understand how the reality of space was transformed in these processes without the focus on cartographic theory and practice.

The issue of positions and disciplines

While emphasizing the significance of cartography, this book offers an approach to the study of space, the understanding of territory and the relationship between territorial space and globalization that is interdisciplinary and evades usual methodological positioning. The argument that modern political territory is cartographic rather than physical or natural fits well into the stream of post-structuralist-inspired writings. Yet, the approach followed departs from postmodern writings in quite significant respects. First, I identify a number of shortcomings within postmodern analyses in terms of understanding space within most writings on territory. The state, and territory, should not necessarily be seen as a logocentric discourse. As will be discussed, it is possible to decentre the modern map, and in consequence, it is possible to inscribe new meanings and centre the modern map differently. Hence, territory does not necessarily have a fixed meaning as it is often claimed in those critical interventions. Second, and as mentioned above, my argument departs from much post-structuralist analysis by taking issue with their predominant notion of globalization. There is much focus on mobility and flows, but less on the apparent solidity of space. Too often a fixed world of territorial spaces is juxtaposed with a global dynamic space of flows and networks. A central implication of this book is that modern territoriality is based on a social network that made space mobile before it could be turned into fixed territory. This serves to challenge both the alleged contradiction between territory and network (for a similar argument, see Painter 2006), as well as that between territorial and global space.

While my argument concerning the reality of space borrows significantly from post-structuralist ideas on knowledge, it departs from these when it comes to understanding the identity of the state, the globe and possibilities for the future organization of politics. Rather than looking at these issues through the lens of textual analysis, I believe that a focus on the agency involved in the historical transformations of the state and the globe provides a more adequate perspective. Yet whereas Historical Sociology has been prolific in historicizing current political structures, it shows a profound lack of engagement with the historical production of knowledge; which is what I, in this book, call the establishment of reality.[2] In consequence, Historical Sociology has its strengths exactly where post-structuralist approaches have their weaknesses and vice versa. Simply speaking, Historical Sociology excels in historicizing the political structures, whereas post-structuralism excels in historicizing knowledge.

A deep division is usually considered to exist between post-structuralist writings and more rationalist approaches. This is often framed as a question over the possibility and desirability of a 'Social Science', where Max Weber is quoted as an inaugurator of a modern value-free science and Michel Foucault is positioned as pioneering a post-structuralist critique of such a thing (for a good example, see Smith 2004). Others have suggested that there is as much continuity from Weber to Foucault as there is rupture; that that they work with strikingly similar themes but 'associated with opposing corners in the contemporary intellectual landscape' (Szakolczai 1998: 1). The perceived division has much to do with the general purpose of science, and the belief in reality, that is, whether there is a reality out there independently of theory or not. Latour (1999) has characterized this dispute between approaches conceptually based on 'reality' and 'language' as 'science wars' and he argues that the division is misguided. There are no separate 'spheres' nor a stable independent reality, but, he suggests, this should not prevent us from talking about reality. Latour's science studies are clearly influenced by post-structuralism; as I will discuss in subsequent chapters he draws on philosophical pragmatism and his notion of a contingent reality bears heavily on post-structuralist ideas. Yet he departs from the typical textual analysis of knowledge production, and investigates, instead, the specific agency involved in knowledge production. It is thus possible to study the establishment of a spatial reality by investigating the agency involved herein.

This book, then, is preoccupied with a problem rather than with positions or methodological orthodoxy. The approach is interdisciplinary and an important theoretical aim is to bring attention to an area of study (cartography), which is largely ignored in conventional IR and globalization theory, in order to provide an answer to questions that are not solved within a narrow understanding of disciplines. In the effort to navigate between Political Geography, Sociology and IR, I am presenting each of these literatures in a way that does not assume familiarity from the reader. While this makes the book, to some tastes, overtly theory-heavy, the aim has been that one could read the book without being familiar with arguments that could have been taken for granted had the book been written from a decidedly IR, or Political Geography, perspective. Historically, the argument will follow the convention that the world was mapped in its present form largely by Europeans. Hence the predominant focus on the development of cartography in Europe and the focus on European practices. This should not in any way be taken as an argument that the technologies involved were endogenous to Europe or that Europe constituted a clearly demarcated area representing a unique culture (see Bhambra 2007) or that these practices were a one-way imposition of a European phenomenon on the rest of the world.[3] To state that reality is a social construction is not to say that we can do what we want. To understand the political possibilities and limitations of our time, we must understand how reality was constructed historically.

Breakdown of the argument

To begin, Chapter 2 will identify the specific space problematique with which this book is concerned. It will be established through a reading of those literatures concerned with spatiality of the state and global change focusing on globalization theory, social constructivism, post-structuralism and political geography. I argue these writings struggle to conceptualize space adequately in between 'transformation' and 'the state'. Even though the case is made for a theoretical awareness of space, most influential writings remain stranded in a duality of objective/subjective conceptions of space, an understanding of historical transition positing a divide between modern and postmodern organization of space, as well as a misperceived opposition between 'territory' and 'network' organization of space. I suggest that because of these three problems the literature fails to appreciate the nature and status of the spatial abstraction invoked by cartographic means that seems so crucial for the understanding of territory.

Responding to this challenge methodologically, Chapter 3 is concerned with positing a framework that seeks to avoid these distinctions. Drawing on Latour's science studies and sociology I argue for the usefulness of maintaining the notion of reality. Discussing the reality of space does not posit it as an extra-social phenomenon, but neither should it be seen as 'pure' social construction devoid of objects. 'Objects have agency too' is a common refrain of Latourian studies and this points to how 'humans' and 'nonhumans' enter into specific relations and through this establish a specific reality. The Latourian framework allows me to analyse the history of cartography as a history of establishing different spatial realities. While being helpful in the study of cartography, however, Latour's writings lack a conceptual understanding of the state. The chapter, therefore, aims to establish a conceptual model that facilitates an analysis of both space and the state as historically contingent phenomena. While a specific mode of establishing a spatial reality conditions possible forms of organizing the state, the processes of state formation also condition and affect the production of space.

Chapter 4 narrows the analytical scope by introducing and discussing the history of cartography. It provides a more specific framework for understanding the power of cartography which governs the historical analysis of the following chapters of the book. The chapter draws, generally, on 'critical historical cartography', and the works of Harley more specifically. It makes a distinction between two types of power relations involved in the study: epistemic power and the power of authorship. On the one hand, the notion of epistemic power points our attention to the overall principles for how maps are produced and recognized as being maps. It is a notion that resembles Thomas Kuhn's notion of paradigms, and in sequence, Foucault's work on knowledge production. On the other hand, the power of authorship directs our analysis towards a specific study of the agency of map-making: who does it, under whose directions and so forth. Emphasizing epistemic power in this particular chapter I highlight the transition from a medieval to a so-called modern, or geometric, mode of map-making and I discuss how the new episteme of map-making affects the establishment of a new spatial reality. Illustrating the consequences of this transition the chapter addresses Baudrillard's statement that under the postmodern condition 'the map no longer precedes the territory'. The point is that under the geometric map episteme, the map has always preceded the territory due to the way in which this cartographic mode abstracts space from its immediate social reality.

Moving to a historically informed analysis, Chapter 5 demonstrates how the world was created as a unified stage for political action through cartographic practices. The imperial contest between Spain and Portugal made reference to the world by cartographic means. This precedence led to a vast emphasis on cartographic development and, by analysing the Spanish master map *Padron Real*, it is illustrated how geometric cartography established a new global spatial reality which became the reference for international politics. During the sixteenth century cosmography became a key intellectual discipline which developed the new geographic vision of a unified globe. The actual cartographic practices slowly realized the global image as a *de facto* single space providing the frame, or the stage, for European power politics between consolidating territorial states. Being established in Europe, this new global space was slowly spreading to the rest of the world which had previously been split between different modes of mapping in, for example, China and the Americas. What was peculiar about the 'geometric representation of space' was that it dissolved the symbolic centre of authority characterizing other map regimes, and in that respect was combinable with other cultural traditions. This obscured the way in which it still served the interest of mainly European actors, and it was through colonialism that the European map was gradually universally accepted as the true way of representing space.

The establishment of a universal global space should, crucially, not be seen in opposition to the formation of particular states. Chapter 6 analyses how the cartographic practices altered the conditions under which state rulers identified and developed their territory. Through a historical exposition of the mapping of Danish territory and state formation, it is argued that the cartographic unification of Danish territory preceded the introduction of absolutism and enabled a new way of knowing, and thus developing and controlling, the territory which was necessary for the notion of modern territory to materialize. From the sixteenth century onwards, cartography became a more and more significant tool of state planning and, both conceptually and in practice, the state began to be identified according to the cartographic image of its territory.

The focus on cartography contributes to the discussion of the political organization of space. I argue that modern territorial space is best understood as being cartographic in nature, and this explains the persistence of territorial space in the face of mounting 'challenges' inherent in the globalization thesis. This is so because if the spatial underpinning of territory is cartographic in nature rather than referring to an external physical reality, then the declining significance of distance and geographical space as an obstacle to socio-economic practice cannot be immediately translated into a statement about the conditions for territory. If the reality of space is established through cartographic practice, then modes of cartography present a set of conditions which affect possible ways for the political organization of space.

2
The State of Territory

There was a certain sense of harmonious correspondence between a world of sovereign nation states and the cold war. The spatial image of the state seemed a perfect match with the spatiality of the world. To the extent that it was theorized at all, territorial space was implicitly conceptualized as a billiard ball, as a solid unit interacting with other units according to the mechanical physics of Newton. One ball moves, hits the neighbouring one and thus causes a reaction. The right policy within this world ought to be one pursuing a balance of power seeking to prevent the movement of any ball, and thus preserve a stable system. The main lines of conflict were supposedly those between territorially defined states constituting a system whose image was a collection of different-coloured territories projected on to the map of the world. From one side, the aim was to contain and prevent further spread of the opposite colour. Territorial exclusivity was the rule of the game. The enemy was kept at bay through containment and wall building.

As the cold war came to an end, 'globalization' and the 'global village' became buzzwords of the day (Scholte 1996) and identity politics took on a prime place on the world political agenda. A number of 'ethnic conflicts', not least in former Yugoslavia and in Central Africa, dominated the security agenda in the West. At the same time, the generally perceived sense of change resonated less and less with the timeless qualities of a world order based exclusively on the nation state. The perception of change fed into a more thorough investigation of political identity and the nature of the state, where the territorial expression of the nation state was seen as an often malevolent static identity marginalizing difference and the fluid nature of identity (Campbell 1998). These discussions reverberated through IR as well as Sociology and Political Geography. Approaches drawing on political economy

21

challenged the container image of the state, and the central concept of identity was employed by poststructuralist and social constructivist challenges to the established conception of the state (Taylor 1994).

While bringing the concept of territory on to the agenda, these concerns have largely, and peculiarly, abstained from a discussion about spatiality and what constitutes these different spatialities. As will be shown, most of the accounts challenging 'the Westphalian image' of politics share a reluctance to explicitly engage with the concept of space. While the state generally remains important, there is a sense in which, as a concept, it is caught between too much reification and too much critique. In most of the so-called mainstream 'scientific' approaches the state(-space) is taken for granted as a significant presence which it is not necessary to scrutinize, that is, a reification of state(-space). Among the critics, the ambition is generally to show that other forces, institutions, spaces, etc. are important and, by implication, the state receives less attention as an analytical concept than as an object of criticism. In consequence, it is often quite a simple notion of the state which is criticized, deemed irrelevant or ethically undesirable; and, in a peculiar way, this critique helps to solidify the image of the state which it aims to challenge. As Bartelson argues, 'ultimately, criticism shares the condition of possibility of its object' (2001: 184). If there was no strong state(-space), there would be no reason to criticize it.

The purpose of this chapter is to pinpoint this problem of neglecting an explicit engagement with 'space' despite its central position concerning questions of the state and globalization. This concern brings together the key concepts of territory and globalization, and while a mutual rapprochement is visible between disciplines, as sociologists have become more concerned with issues of space and geographers with the social, seen from an IR perspective, there is still a significant gap to be bridged in terms of bringing these concerns into the core of what constitutes IR. In the discussions, for example, concerning the most important unit, or focus, of analysis in IR, space too often remains an unspoken assumption. In the discussion of whether it is the state, the market or institutions that should attract our attention, the state is typically either reified as being 'the terminal entity that serves as the foundation of [the] discipline [IR]', to quote a paper by James Rosenau (2004), or it is given a significant but not privileged position alongside other polities, as in Ferguson and Mansbach's (2004) *Remapping of Global Politics*.

Rather than emphasizing the diminishing 'capacity' or 'role' of the state in times of transition, claims to globalization ought to provoke more thorough analyses of what the state is, its genealogy and what international relations could possibly be with, or without, it. There has been no shortage of challenges to, and arguments over, sovereignty, but much less discussion of what sovereignty refers to: that is, the question 'sovereignty over what' (Elden 2005a & 2005b)? In order to demonstrate that even those that challenge a reliance of an unproblematic notion of sovereign territory, space is, more often than not, assumed rather than investigated. First, the body of globalization literature mainly associated with the discipline of Sociology, maintains the global as a different kind of space. Second, constructivist contributions have been sensitive to the historical contingency of the state but rely too heavily on meaning giving practices and leave the 'materiality' of space outside what is constructed. Third, the spatial order of international theory and practice has been challenged by poststructuralists arguing that the territorial state has lost its legitimate position, both analytically and ethically. But such writings have abstracted space to the extent that they have been better at 'spacing knowledge' rather than provide 'knowledge of space'. Fourth, and lastly, the reinvigoration of Political Geography has picked up these questions in a manner where space is seen as a social production or construction. Yet, again, I take issue with the way in which space is made social in these writings. Rather than seeing space as an outcome of political, economic or cultural practices on a general level, I suggest to ascribe a notion of analytical autonomy to space in order to investigate how space itself conditions other political, economic or cultural practices.

The central pivot of the reading of this problematique is what Soja has termed 'the political organization of space' by which he refers to the way space, and human interaction in space, are structured to fulfil political functions (1971: 1). Such an approach suggests historicizing space itself in the attempt to understand change. This is to say that the political organization of space is both the outcome of 'how space is produced' and 'how this space is organized'. To begin, I will present a discussion of 'the significance of territory' (Gottmann 1973); why it is important and what is at stake by claiming changes to it. Subsequently I identify three key problems in the literature, bearing on a notion of globalization and/ or spatial change: first, there is a persistent subject-object dichotomy in the conceptualization of space. Second, there is a problematic notion of history positing a dichotomy between a modern era dominated by the sovereign state, now giving place to a late-/ post-modern era where global space is superseding territory. And third, there is a persistent insistence on posing network and territory as opposites reflecting mobility and stasis. While several of the authors have warned against some of these problems, few have addressed these in combination. The three problems point forward to the remainder of the book which seeks to overcome these divides between subject/object, modern/postmodern and network/territory when it comes to the understanding of the political organization of space.

The significance of territory

In two excellent textbooks introducing the concept of territory, David Delaney (2005) and David Storey (2001) present territory as a general dimension of social life that is not exclusively associated with the state. Indeed, Delaney presents territoriality as something which – like language – might be a human universal (2005: 10). As such they follow the ideas of Robert Sack who described territoriality as a 'geographic strategy to control people and things by controlling area' (1986: 5). While drawing on Sack, Delaney broadens the understanding of territoriality beyond a strategy for the control of space. 'It is better understood as implicating and being implicated in ways of thinking, acting, and being in the world – ways of world-making informed by beliefs, desires, and culturally and historically contingent ways of knowing' (Delaney 2005: 12). In that sense, territory is central to our understanding of social order and how power is present in the material world, and Delaney conceptualizes it as an outcome of social space, meaning and power (2005: 13–8). Delaney maintains that territoriality is much more than a strategy for the control of space. 'It is better understood as implicating and being implicated in ways of thinking, acting, and being in the world – ways of world-making informed by beliefs, desires, and culturally and historically contingent ways of knowing. "Territory, in turn, informs key aspects of collective and individual identities"' (Delaney 2005: 12).

While it is difficult to distinguish the 'spatial' from the 'social' in Delaney's approach to territory, there is a degree to which he downplays the 'thingness' of territorial space. And by emphasizing territorializing practices he 'situates territory more firmly within the realm of social action' (Delaney 2005: 16). What I would like to add to such an understanding of territory is that there are practices of space making that are more general than those territorializing practices highlighted by Delaney. These concern a particular way of establishing a reality of space, a particular view of the world and they are significant because if 'space' precedes 'territory', then the assemblage of a particular spatial reality conditions possible territorializing practices. And this is particularly important in the context of the spatial identity of the state and the globalizing practices of the European renaissance. Whereas it is essential to remember that territory in itself is neither modern, Western or exclusively linked to the state, then it is the particular link between sovereignty and territory that contributes to the specificity of the so-called modern state. And to understand this, we must not only think in terms of territorializing the state and sovereignty but investigate a broader notion of spatial change that includes a notion of globalization preceding or co-occurring with the formation of territorial states.

Focusing on the relationship, then, between territory and the state, it is useful to be reminded of Jean Gottmann's seminal work *The Significance of Territory*, where 'territory appears as a material, spatial notion establishing essential links between politics, people, and the natural setting' (1973: ix). In his view, territory is significant 'as the unit in the political organization of space that defines, at least for a time, the relationships between the community and its habitat on one hand, and between the community and its neighbors on the other' (Gottmann 1973: ix). The link between territory and sovereignty is essential but also one that is changing over time as 'the basis for the enforcement of the law subtly shifted from allegiance to a personal sovereign toward controls exercised by the sovereign power in geographical space. The partitioning of space thus acquired an increasing significance, and territorial sovereignty became an essential expression of the law coinciding with effective jurisdiction' (Gottmann 1973: 4). The significance of territory, then, lies in the spatialization of state sovereignty that served as a basis for the conceptualization of international politics as something taking place between spatially differentiated but similar (in that they are sovereign) entities. This has been the self-understanding of the IR discipline, where the peace treaties of Westphalia are seen as the foundation of modern IR (Walker 1993). And it is important to recognize this while at the same time remembering that territory has much wider meaning and the understanding of political space should not be limited to simply territory. The problem for IR, as I will show, rests in its failure to discuss territory as a spatial issue, and hereby overlooking a significant condition for the territorialization of sovereignty. In his seminal critique of the spatial assumptions in IR theory, John Agnew suggested that '[i]t has been the geographical division of the world into mutually exclusive territorial states that has served to define the field of study [IR]' (Agnew and Corbridge 1995: 78). Yet, at the same time, these assumptions have been left unexamined and rather than being theorized as a central concept space plays a marginal role in most IR theory. This has particularly been the case after the scientific aspirations of behaviouralist and quantitative methods became dominant in the discipline. Prior to that, however, Political Geography and IR share a common heritage in the geopolitical tradition building on a similar spatialized image of the state.

The geopolitical tradition

The term 'geopolitics' was famously coined by the Swede Rudolph Kjellén in 1899 for the study of the relationship between geography and the character of states as living organisms (Tunander 2001). Central to geopolitics was the idea of the state which, as a natural organism, required an ever increasing space in order to accommodate an increasing population (Bassin 1987: 477, n.7) as well as the deduction of the political from natural spatial determinants (Teschke 2006b: 327). In Sven Holdar's interpretation, Kjellén adopted the organic analogy of the state primarily to visualise the view of the international system as anarchic (1992: 319). Kjéllen's work was inspired by Friedrich Ratzel's political geography, where territory was seen as an expression of the dynamic power of a state. He employed a universal logic in the attempt to establish a science of political geography entailing unity of all organic life where human society was thought to be governed by the same laws as those governing the natural world (Bassin 1987: 477). In Ratzel's own words,

> The territory of a state is no definite area fixed for all time – for a state is a living organism, and therefore cannot be contained within rigid limits – being dependent for its form and greatness on its inhabitants, in whose movements, outwardly exhibited especially in territorial growth or contraction, it participates. Political geography regards each people as a living body extending over a portion of the Earth's surface, and separated from other similar bodies by imaginary boundaries or unoccupied tracts.
> (Ratzel 1896: 351, also quoted by Ó Tuathail 1996: 37)

The more advanced the state is, writes Ratzel, the greater the appreciation of the significance of the soil (1896: 358), and as 'the appreciation of the political value of land becomes greater, territory becomes to a greater degree the measure of political strength and the prize towards which the efforts of a state are directed' (Ratzel 1896: 360).

He famously coined the term *Lebensraum* designating the need for nations to have space to grow and live, thus legitimizing territorial expansion of vibrant states (Ó Tuathail 1996: 37–8). As such he posited existence as a struggle for space:

> Between the movement of life, which is never at rest, and the earth's territory (Raum), which remains constant, arises a contradiction. Out of this contradiction the struggle for space is born. [In the beginning] life was quickly able to [spread and] take over the land surface (Boden) of the earth as its own, but when it reached the limits of this surface it flowed back, and since this time, over the entire earth, life struggles with life unceasingly for space. The much misused, and even more misunderstood expression 'the struggle for existence' really means first of all a struggle for space. For space is the very first condition of life, in terms of which all other conditions are measured, above all sustenance.
>
> (Ratzel quoted in Bassin 1987: 479)

This quote not only captures the central tenet of mixing social Darwinism with a logic equating the size of the territory with power, it also presents an interesting distinction between life in constant movement, and space which is a constant and static. And he seems implicitly to commend those civilizations that are mobile as superior whereas the agriculturalist societies are stagnant (Ratzel 1896: 359).

Continuing this tradition with the focus on world domination, Halford Mackinder argued that the control of the Eurasian heartland was the key to command the world. In *The Geographical Pivot of History* he argued that the world had entered 'the post-Columbian age' which denoted a global and closed political system. This meant that there was no more 'open space' left, and hence, all social action would from then on 'be sharply re-echoed from the far side of the globe' (Mackinder 1904: 422). The political power of competing peoples was a product of their geographical condition and 'the relative number, virility, equipment, and organization of the competing peoples' (Mackinder 1904: 437). The other issue worth noticing is his focus on how technology manipulates the spatial conditions. The core of the argument is that the expansion of railroads was shifting the balance towards land power (Mackinder 1904: 434). Because of the ability to integrate the great landmasses of Eurasia, and improved mobility, Russia would replace the Mongol Empire as a superior neighbour. Europe had historically been contained by the superior mobility of Asian 'horsemen'. This power relation changed when improved sailing technology increased sea-power vis-à-vis land power, and now (at the time of writing) the relationship was changing again due to increased mobility over land.

While the association with Nazism undermined the position of geopolitics as a legitimate subject, the ideas maintained a significant place in strategic studies (see, for example Gray 1988: 4; Gray 1977), and the relationship between space, the state and international relations was discussed by authors such as Raymon Aron. In *Peace and War* he sought to provide a critical reflection on 'the nature and limits of geopolitics' (Aron 1966: 182), because since the international order was based on territorial sovereignty, it was necessary to grasp political geography in the study of IR. Explicitly continuing Mackinder's geopolitics, he argued that space, along with number (population) and resources, 'define the causes or the material means of a policy' (Aron 1966: 180). Arguing that space was both the environment, theatre and a stake of IR, Aron suggested that space is never simply causal in itself, and to understand the spatial environment we must understand it as a specific historical conjunction. Interestingly, Aron picked up a notion of 'sense of space' from Carl Schmitt which emphasizes how different intersubjective conceptions of space exist in different settings: '[t]he sense of space has been, in each period, determined by the image which men have made for themselves of their habitat, by the style of movement and combat on land and at sea, by the stake for which societies come into conflict' (Aron 1966: 207). In that respect Aron distanced himself from 'the geographical causation in universal history' presented by Mackinder (Aron 1966: 197); yet, he also dismissed the utopian dreams of the liberal tradition. The fact that space was no longer a natural constraint, due to technological development, does not mean that conflict over territory will not occur.

We also find the topic of how the international system was based on a spatial organization of sovereignty in the work of classic authors, such as E. H. Carr and Hedley Bull. Carr argued that territory plays a crucial role for what the state is, and territory is important, in that it is a source of military power, which is seen as an inescapable aspect of politics. However, Carr notes that territorial organization will not necessarily always be the foundation of political power. In sequence, he asks whether the largest and most comprehensive units of political power in the world necessarily are of a territorial character, and if so, whether they will continue to take approximately the form of the contemporary nation state (Carr 1981: 210). Carr foresaw a tendency driven by the development of communications technology and expanding capitalism in which the units of political power would grow bigger. Instead of having political power distributed between equal sovereign states,

Carr contended that we are likely to see larger units of authority composed of several formally sovereign states, acting together around one single centre of authority (1981: 210–13). Following this theme, Bull famously introduced the notion of 'new mediaevalism' (1995: 245–6) in his discussion of alternatives to the international states system. It denoted 'a modern and secular equivalent of the kind of universal political organization that existed in Western Christendom in the Middle Ages' (Bull 1995: 245) representing a system of overlapping authority and multiple loyalties. The most obvious example of this is the general trend towards regional integration as exemplified by the EU (or, as it was then, EC) process (Bull 1995: 255).[1]

Hence, these authors assert the significance of the space-sovereignty relation but they also challenged the organic connection between territory and sovereignty as providing the foundation for the identity of the state. And while they operate with a somewhat simplistic conception of space, they present a research agenda that one could expect to gain ground within IR dealing with an increasingly complex international system. Yet the contrary has been the case, particularly after the scientific aspirations of behaviouralist and quantitative methods became dominant in the discipline. Intriguingly, this meant the disappearance of a concept, which had provided the foundation of a geopolitical science prior to the Second World War, as a Political Science with positivist and universal aspirations gained ground. In Waltz' classical formulation of Neorealism neither space nor territory featured in the system. Echoing Weber's famous definition of the state, it is sovereignty that defines states and constitutes them as like units, yet in a manner symptomatic of rationalist IR's neglect of space, there is no reference to what sovereignty is over, what it refers to and what limits it. Not even when Waltz argues that domestic governments have a monopoly on the legitimate use of force are 'territory' or 'boundaries' mentioned (1979: 93–7). Widening this critique beyond Waltz, the prevalence of rational choice-informed IR theory has extended what, echoing Agnew's critique of Neorealism (Agnew 1994), could be called a conceptual entrapment of space where it appears as an implicit assumption that is never discussed. This refers to the fact that assumptions about territorial space have been crucial for IR theory at the same time as remaining hidden, or entrapped, in the theoretical framework which made space near impossible to talk about.

After the fall of the Berlin wall, discussions of space began to re-emerge on the research agenda although most IR still did not fully recognize its significance. There was a growing territorial dissidence questioning the predominant rationalist paradigms of most social sciences at the same time as events in the world seemed less and less in harmony with an imagined world of fixed spatial divisions. In particular, the assumptions about territory informing rationalist theory were questioned, as I will discuss subsequently. In the words of Gearóid Ó Tuathail: '[t]erritory [...] is no longer the stable and unquestioned actuality it once was. Rather than assumed given, its position and status is now in question' (2000: 139). Hence, the 'fall of the wall' can be seen as symbolic of an unlocking of a closed spatial system both in politics and in IR theory. Correspondingly, a significant body of literature on globalization has fed into and complemented the arguments which have put spatial issues on the IR agenda.

A new global spatiality

The theoretical globalization literature tends to depend on claims to spatial and temporal transformations (Rosenberg 2000). In *The Production of Space*, Lefebvre (1991) stresses how space is produced and is always political and ideological. Instead of looking at how a notion of 'natural space' – or, a space which is simply there – relates to human agency, Lefebvre turns the attention to how all space we live in is social space and is a product of political and economic processes. By implication, global space should not be regarded as a natural space and so to talk about global space would also entail a socio-political production. In other words, such a perspective alerts us to the extent that not only territorially organized states are political spaces, but also how 'a global space' must invariably be a result of socio-political processes. As such territory and global space should not, instinctively, be perceived as opposites but related. This however, has not been taken up by the most common approaches to globalization. Especially in the 1990s it was becoming a widespread belief that the state, and (territorial) space, would gradually lose significance and vanish in face of the forces of globalization. Fuelled by an economic logic, Richard O'Brien (1992) famously put forward this claim (see also Graham 1998). In a similar vein, Kenichi Ohmae (1995) proposed that we were seeing the end of the nation state and instead the rise of regional economics.

Consequently, such authors reproduce an old debate between geopoliticians or nationalists and cosmopolitanists or liberals. This is a discussion that can be traced back to enlightenment thinking about the role of the nature of the terrain for the development of society, and environmental explanations for the variance in constitutions, laws and customs that can be observed between different countries (for example,

Montesquieu 1995; see also Heffernan 1999). Before the first World War, Norman Angell articulated a profound critique of the widespread idea that military and political power gives a nation a commercial advantage (1910: vii) and further states that 'the idea that addition of territory adds to a nation's wealth is an optical illusion of like nature, since the wealth of conquered territory remains in the hands of the population of such a territory' (Angell 1910: vii). Apart from opposing the idea that there can be any economic benefit from war and territorial expansion, he suggested that the state generally was 'from the ancient' (Angell 1910: 225); that due to developments in communication and technology and the growing significance of internationally organized class and capital, what we have in front of us is 'the organization of society on other than territorial and national divisions' (Angell 1910: 253). In the context of globalization theory, it is striking how his argument rests on a notion of how the 'rapidity of communication creating a greater complexity and delicacy of the credit system'. With catchy statements, such as '[t]he capitalist has no country' (1910: 248), much of Angell's argumentation reverberates around similar themes to those put forward in current globalization accounts and the similarity is only underlined by Angell's complaint that our ideas about international politics 'are still dominated by the principles and axioms and terminology of the old' (both from Angell 1910: viii–ix).

The title of the book, *The Great Illusion*, referred to the idea that it was necessary to maintain large defence forces to secure the prosperity of nations (Angell 1910: 24) but after the first World War, where the organized working class had not prevented the war, and nationalism had not been vanishing as a social force, it was easy to turn the title back on Angell's own argument. In fact this was already done in a review of the book soon after publication by another of the great geopolitics writers, Alfred T. Mahan, who suggested that Angell's work was indeed an illusion 'based upon a profound misreading of human action' (Mahan 1912: 332). What is at stake in these arguments concern the necessary relationship between space and society; and these early contributions to an emerging IR literature, not yet institutionalized, also present dilemmas that reverberates through the current discussions about globalization and the significance of territory (Knutsen 2007). In this light there is nothing new in the theoretical discussion about globalization. However, the radical opposition between state vis-à-vis a global market posited in these writings has now been more nuanced where the focus is more on the changes that the state is undergoing (see, for example Brenner 1999). And while this literature is becoming increasingly diverse and nuanced, there is a continuing tendency to make the tension between, and possible change from, territorial space and/to globality the core feature of globalization theory.

One of the standard references, Jan Aart Scholte's *Globalization*, presents an explicit attempt to rethink space, territory and change as central to globalization. Scholte states that the challenge for social research is to examine the intricate interplay between what he calls 'globality' (flows in a unified space) and 'territoriality' (fragmentation) (2000: 60).[2] Hereby, he posits what seems to be a problematic distinction between territoriality (fragmentation) as continuity and globality (unification) as change. Scholte's main argument is that we are living in a time of change which necessitates new analytical concepts. He then coins the catchy term 'methodological territorialism', which refers to 'the practice of understanding the social world and conducting studies about it through the lens of territorial geography' (Scholte 2000: 56), to describe mainstream conceptions of society. And these are no longer adequate because an increasing number of social activities now take place in a distance-less space of globality. Such social relations transcends the territorial spaces hitherto the foundation for social science. Scholte therefore argues that globalization should be understood as deterritorialization equated with supra-territoriality. By this he means that what we see in the world today is a reconfiguration of geography 'so that social space is no longer wholly mapped in terms of territorial places, territorial distances and territorial borders' (Scholte 2000: 16). Indeed, Scholte states that we are seeing a far-reaching change in the nature of social space (2000: 43).

Before the processes of globality gained momentum – which they did during the 1960s according to Scholte – a territorial geography was apt to describe social relations. Now, however, we must rethink geography: 'In territorial geography, relations between people are mapped on the earth's surface and measured on a three dimensional grid of longitude, latitude and altitude' (Scholte 2000: 47). And, argues Scholte, the process of globality implies that we must change the map of social relations and move to operate a four-dimensional space (longitude, latitude and altitude + globality) (2000: 61). He describes this as an identifiable shift in ontology caused by globality which involves a changed perception of space from 'territorialism' to 'supraterritorial space'. In other words, we can no longer comprehend world geography in terms of territorial spaces alone, Scholte argues, and therefore globalization requires us to substantially rethink global theory (2000: 315). Despite the immediate logic of these arguments, Scholte has actually provided two very different notions of what space is, and more importantly, a problematic history of political space.

Firstly, Scholte invokes a solid subject/object distinction in his understanding of space. The question for Scholte is how to capture the fact that social relations not only occur in space, they also produce space. On the one hand, space is something that is just there, as an objective fact, to which social relations relate in varying ways. In this understanding space remains a stable object and changing social relations are caused by globalization. As he states, '[g]eography ranks on par with culture, ecology, economy, politics and psychology as a core determinant of social life' (Scholte 2000: 46). Here, space becomes something distinct from the other dimensions such as politics, culture and ecology, and it is not quite clear how they relate to space in the sense that these will all have a spatial dimension. Now, on the one hand, for Scholte space refers to some physical attributes that surround us, hence, we see difference in lifestyles between villagers living in mountains, and villagers living by the sea (space as objective reality). Yet, in other instances, space is something that is affected by the activities going on in these spaces (space as a social product). The tension between space as a natural 'background' variable and a social product remains unresolved exactly because he relies on a subject/object distinction (nature/social) in his attempt to reach new theoretical models to capture globality.

Secondly, in his effort to make globalization something qualitatively different from the past, Scholte invokes the common, yet troubled, distinction that there used to be a territorial geography, but now we have added a 'special' postmodern dimension which transcends the way of territory. A significant question here is whether what Scholte called a territorial geography was ever apt for describing social relations? And if it really was, it would be significant to know how this came into being in order to know whether it is really being transcended by globality because 'territorial geography' was never simply there, but was a very specific political production, or representation, of space. Admittedly, Scholte issues a warning not to write off the old world of territorialism. Nevertheless, if globalization invokes such a radical change that we should talk about a new ontology of space, we certainly also need a more specific account of the history of space and of the state.

In consequence, and thirdly, he posits the territorial logic of modernity as being qualitatively different than the network logic associated with the postmodern. This resonates well with another influential account on globalization; in his three-volume opus *The Information Age* Manuel Castells (2000) ascribes a central role to space and to the idea of the emerging 'network society'. He sees space as the expression (and not reflection) of society, and for him it is necessary to identify the historical specificity of social spatial practices (Castells 2000: 441). There are new spatial forms emerging, and a new logic of space which he describes as 'space of flows' defined as 'the material organization of time-sharing social practices that work through flows' (Castells 2000: 442). Castells' argument is not that the network society is made up entirely of this new spatiality, but rather that there is a tension between the majority who live in 'space of places' with diverse temporalities on the one hand, and a minority of managerial elites living within the different logic of 'space of flows' on the other. Hence, these elites become more and more isolated from the rest of society – in a somewhat abridged expression Castells states that elites are cosmopolitan (remember, '[t]he capitalist has no country' (Angell 1910: 248)) and people are local (Castells 2000: 446 & 499).

Castells' 'space of flows' resembles what Scholte adds as a fourth dimension of globality to the three-dimensional geography of modernity in, what I would describe as, a 3+1 model of global space. And even though globalization, as Ian Clark tells us, essentially represents an attack on the great divide between inside and outside, domestic and international politics, as distinct realms of activity (1999: 10–18), accounts such as Scholte's and Castells' tends to reinsert divides. However, as the example of Scholte showed, without a specific notion of space, and how space is historically produced in various settings, such an argument easily fails to deliver what is promised and instead reproduces simple distinctions between natural/social space, between a static territorial geography of the past/distanceless space of speed for the future, and the network vs. territory logic ascribed to different eras. Although Castells employs a 'proper' theory of space, neither he nor Scholte present the reader with a convincing account that captures both the unifying trends of globalization as well as the fragmentation of this unitary space into disparate political entities. Instead, they emphasize the production of flows. And, as Clark poignantly suggests, the view of globalization as transcending the territorial space of the state misrepresents the nature of the processes and neglects the role of the state within these (1999: 39). And in both accounts, the state is peculiarly absent as anything else then a stable background category, a leftover from the past which lives an antagonistic life in the three dimensional geography of the 'space of places'.

In more general terms, Hannes Lacher has identified two main problems associated with globalization theory: firstly, a failure to capture the dynamics of interstate competition and geopolitical pressure in the processes of globalization; secondly, a flawed understanding of history indicating a transition from the national-territorial modernity to a global postmodern era. In his words, '[g]lobalization theory has indeed failed to develop an adequate conceptualization of the role of the state [...] and of interstate competition in the processes of socio-spatial restructuring that have taken place over the past quarter century' (Lacher 2006: 4). Lacher, thus, concludes that 'globalization theory, with its notion of a transition from a national to a global age, must remain an unsatisfactory intellectual framework for the consideration of the socio-spatial transformations of the present' (2006: 5). Lacher would like to maintain that since capitalism was never contained by state units, such a territorial geography as promoted by Scholte, was never an apt description of modernity. What is needed instead, Lacher argues, is a proper 'study of the social production of political space', and this involves contending that the sovereign state was ever a true master of space, even in the modern period (2006: 15). Nevertheless, where Lacher's general diagnosis of the (spatial) globalization debate in IR is pertinent, I will suggest, and later discuss, why I think he overemphasizes 'social practices' in the production of different spaces, and refrains from any significant discussion regarding the concept of space itself; how it relates to social practice, and how the reality of space in and of itself may change. The predominant subject/object distinction making social space subjective, and natural space objective, is problematic because state territory tends to be something which is either seen as a social space, and thus a historically contingent spatial mode often explained as the outcome of other forces and social practices. Or, it is considered a natural space, which is simply there. In both cases, physical, or geographical space, is made redundant; something we as social researchers do not have to deal with or something that is seen as being overcome by technological progress, spreading of financial markets and so forth. As Doreen Massey argues, the idea of space as a surface on which social practice occurs is 'taming the challenge that the inherent spatiality of the world presents' (2005: 7).

Social constructivism with spatial history

From a social constructivist viewpoint the sovereign territorial order and possibilities for change have been investigated in a manner that through deconstruction of historical processes seek to point to possible contemporary changes and to a different understanding of 'the international'.

John Ruggie's seminal article 'Territoriality and Beyond...' (1993) was one of the first non-radical challengers to the spatial order of

Neorealism's timeless model of world politics in IR (see also Kratochwill 1986). Ruggie argued that territorial rule is neither necessarily exclusive nor anarchic, but, on the contrary, there is historical evidence for overlapping authority structures which fit neither anarchy nor hierarchy. This argument was based on a historical interpretation of the origins of territorial rule, and thus, the spatial organization of international politics. Ruggie maintains that the transition from the medieval to the modern political system was an epochal change, and that the 'central attribute of modernity in international politics has been a peculiar and historically unique configuration of territorial space' (1993: 144). In order to understand this, Ruggie states that we must grasp the fundamental change that occurred in the way political space was conceived. The 'demise of the medieval system of rule and the rise of the modern resulted in part from a transformation in social epistemology. Put simply, the mental equipment that people drew upon in imagining and symbolizing forms of political community itself underwent fundamental change' (Ruggie 1993: 157). The key point to this change, was that political space changed from being organized around a variety of authorities, all ultimately subject to the universal realm of god, to being *'defined as it appeared from a single fixed viewpoint'* (Ruggie 1993: 159, italics in original). With these arguments, Ruggie ascertained the historic specificity of the sovereign territorial state, and thus the configuration of the international system.

Centred on global capitalism, Ruggie has defined a new era – a new epoch – where we see an 'unbundling of territory' that signifies a change in political space. He stresses this by stating that: 'Unbundled territoriality is not located some place else; but it is becoming another place' (Ruggie 1993: 174). Following the steps of Hedley Bull, he suggests the EC (now EU) as a 'multiperspectival polity' representing non-territorial modes of rule. The main thrust of Ruggie's article is concerned with establishing the argument concerning an epochal break from the middle ages to the modern, and his prescriptions for late twentieth century change remains suggestive. In more recent work, Ruggie (2004) argues that there is a distinct global public domain emerging and that this represents a new world of transnational spaces constituting a potential historical progressive platform. It is 'an increasingly institutionalized transnational arena of discourse, contestation and action concerning the production of global public goods, involving private as well as public actors' (Ruggie 2004: 503). Making these claims, Ruggie is less concerned with spatial arguments than he is with asserting an emerging specific global system of governance that cannot be reduced to the (inter)nation state politics.

Space, however, takes on a prominent place in Ferguson and Mansbach's *Remapping Global Politics* (2004). Continuing Ruggie's train of thought, they focus on the declining capacity, legitimacy and authority of states as we are entering a new epoch of post-international global politics. The aim of the book is 'to redraw our mental maps of global politics and explain the forces shaping change. Recognition of the changing nature of political space and time takes us well beyond the two-dimensional cartographic representations that were deemed sufficient for centuries' (Ferguson and Mansbach 2004: xi). This quest to redraw our mental maps is informed by a strong notion of transition from 'internationalism' (governance and policy making is rooted in interstate relations) to a 'postinternational world' that 'features an increasingly variegated universe of polities, identities, and loyalties' (Ferguson and Mansbach 2004: 342). We are entering a period where patterns of authority are only going to get messier, where the simple world of nation states is no longer appropriate.

This piece is, in many ways, symptomatic of the approach to space in IR that seeks to find new ways to make sense of global change. While it introduces the terms of (re-)mapping, political space and time, the treatment of space remains on a metaphorical, or a very abstract, level. Being true to the ideational imperative of this strand of constructivism, any notion of materiality, or objectivity, is written out of space. 'Political space refers to the ways in which identities and loyalties among adherents to various polities are distributed and related, and territorial space is only one of the possibilities' (Ferguson and Mansbach 2004: 67). They then argue that political space can be organized in a number of different ways and that technologies are facilitating this reorganization. As the connection with actual land has decreased in importance, 'the concept of space also changed, until today territory remains only one of many ways to define the extent of a polity's domain, that is, the political space that it occupies' (Ferguson and Mansbach 2004: 67). They continuously hold sway to the declining relevance of territory in global politics and for the proliferation of alternative conceptions of political space. This declining role of territory is 'essential to the transformation of the international world into a postinternational one, especially in producing new identities and redefining old ones' (Ferguson and Mansbach 2004: 71).

In sum, I have already questioned the notion of a temporal distinction between a modern IR characterized by a territorial geography, and a post-/late-modern globality characterized by a radical different spatiality. But Ruggie eventually reproduces this divide indeed because he abstains from adequately discussing whether the episteme of space is actually changing now in a way which is comparable to the spatial transformation which took place during the Renaissance. And more importantly, by focusing only on social representations of space he abstains from a more general discussion about the relationship between space and politics. A similar criticism can be raised to *Remapping*, in which the authors are unable to differentiate between territorial space and other types of (functional) political spaces. This also means that their desire to present a historical exposition, shedding light on future changes to the organization of political space, remains unable to grasp differences and similarities between the power of organizing and controlling territorial space and what is going on in transnational networks of corporations. Thus, even though they seek an improved historical understanding in order to grasp the present, and plead for going back in time to a pre-international world to get a greater perspective on the present and the future, they reproduce the main error of Ruggie by not adequately discussing territory as space proper; that is, the spatial foundation of territory. How is space produced in different eras and what allows a transition from the pre- to the post-international?

Ferguson and Mansbach boldly state that maps today are increasingly misleading because they present a world of states as box-containers and this is no longer a true description of the world (2004: 31). Yet, as discussed above, this image might never have been an apt representation of politics even if it has remained prevalent. This means that something else is going on regarding the power of space, and the power of representation. Something that we have to investigate more closely rather than reproducing the notion that we leave a modernist 'territorial geography' behind while entering a late modern globality. A significant flaw in this line of thinking is the crucial question that they do not ask: Why is global space a more natural, or even real, venue of politics than territorial space?

Poststructuralist spatial interventions

From the late 1980s, radical critical writings inspired by poststructuralism generally, and Michel Foucault and Jacques Derrida in particular, started to question established knowledge and conventions in IR. Some of the most important writers have been Richard Ashley, Rob Walker and David Campbell. They challenged the dominant scientific aspirations and often criticized the discipline for being implicitly involved in constituting the world that it was seeking to describe. However, they also challenged the 'corner' from which previous critiques had been articulated: Historical Materialism and various branches of Marxist thought. They were critical of the linear causality and the 'metanarrative' of capitalist development, which informed the bulk of these writings. However, a key focus among these writers has been 'state-centrism' and the overt reliance on the state in IR. A common argument in poststructuralist thinking is that the state has lost not only its analytical legitimacy, but also its ethical legitimacy as the natural locus for the political community. Part of this challenge was an essential questioning of the spatial assumptions informing the discipline, not only in the guise of territory, but also space more generally.

Rob Walker's critical reading of the canon of IR theory, and his argument that modern politics is spatial, has provided an important source of influence for this book. Even so, Walker's writings are lacking in their analytical capacity precisely because he is deconstructing the modernist spatial assumptions of the IR discipline, but has no concept to put in its place. Nevertheless, Walker has presented one of the most compelling deconstructions of international relations, especially with regard to the relationship between space and the political, and one that has provided inspiration outside the field (see for example Agnew 1994 and Sparke 2005). In his *Inside/Outside: International Relations as Political Theory* he is concerned with the study of IR texts as an expression of the limitation of our political imagination. He focuses his analysis on three groups of problems: one is concerned with modern political identities; the second with the understanding of historical change; and the third is concerned with the theoretical critiques associated with postmodernism and poststructuralism (Walker 1993: 11–12).

According to Walker, modern politics is spatial politics relying on a spatial distinction between inside and outside. 'Political life occurs in space. So much is at once both obvious and obscure. For ideas of space, like those of time, express many of the greatest mysteries of human existence' (Walker 1993: 127). He then argues that space has been prioritized over time in modern accounts of international politics. Accounts of plausibility reflect a spatial imagery far more readily than considerations of possibilities in time. Hence, as 'a domestication of an inside and an outside, state sovereignty specifies a politics of spatial containment; one, moreover, that specifies a new range of possibilities or a human existence in time' (Walker 1993: 34). It should be noted here that, curiously, a parallel but reverse criticism has been articulated by Massey, who argues that time has been prioritized over space in modern social theory (Massey 2005: 21; see also Soja 1989). Although this seems to contradict Walker's statement, it actually only reaffirms the fundamental 'inside/outside' distinction that Walker is pointing to; in IR it is the spatial divisions of the world which are thought to be of primary importance, whereas 'inside' society it is temporal progress which has captured the high ground.

The spatial politics of modernity, Walker argues, rests on a Euclidian/ Newtonian conception of space and space-mechanics which also informs much structural explanatory theory. He suggests that the worst caricature of the state is the somewhat familiar image of the billiard ball as it appears in Waltz' Neorealism. Walker, then, traces the historical development of dominant perceptions of space which bear on a mathematical formalization of space in a combination of Newton's distinction 'between absolute homogenous space and the space of ordinary experience which is treated as merely apparent and relative' (1993: 129). The true absolute space of Newton is readily describable by the basic postulates of Euclidian geometry, which, among other things, assumes the homogeneity, limitlessness, continuity and infinitude of space. For Walker, the essential feature among these is the homogeneity of space, which has played a crucial role in determining the cultural forms of European civilization (1993: 129). It was the combination of such conceptions with the reinvention of a single-point perspective in visual arts that coincided with a cartographic 'revolution' that shaped a linkage between the idea of sovereignty and ideas of sharply delineated homogeneous territorial spaces that characterize the so-called modern territoriality.

The empirical analysis of subsequent chapters in this book indeed aims to show how space has been established historically so that it coincides with the ideals of modern space as described by Walker. Yet his notion of change reveals the absence of a general notion of space, and a somewhat simplistic analytical understanding of the state and change. Even though he states explicitly that he will not provide yet another explanatory theory, Walker cannot avoid inducing analytical propositions (1993: 159). He suggests a link between the current experience of accelerating temporality (and 'the annihilation of space by time') with a crisis in, or inadequacy of, the spatiality of modern politics: 'the hope that temporality may be tamed within the territorial spaces of sovereign states alone is visibly evaporating' (Walker 1993: 155). In his usual eloquent style, he states that '[t]he language of probabilities and accelerations now familiar from astrophysics contrasts sharply with the restrained dynamics expressed in the great Newtonian synthesis of cosmic order' (Walker 1993: 5). This is to link the Newtonian concept of absolute space with modernity, and an 'Einsteinian' concept of relative space with postmodernity. By doing this, he opens up a future where it is possible to think about political identity beyond the territorial confines of the sovereign state resting on a notion of absolute space.

By implication, Walker is reproducing the now familiar theme of a spatially fixed modernity whose ideals are becoming increasingly untenable in the face of the temporal acceleration of globalization. Sovereignty can no longer serve as 'a spatial resolution of the relationship between universality and particularity' (Walker 1993: 78), and, in consequence, politics is in an ambivalent state. Yet, the distinction between an absolute space of modernity and a relative space of postmodernity only affirms a troubled distinction and fails to capture the degree to which temporality and spatiality might go hand in hand, rather than being antagonistic opposites. To reinstate Marx' dictum that space is being annihilated by time is to reproduce the flaw of 'domestic' theory, as pointed out by Massey above, to prioritize time over space.

In the writings of Ashley, including joint pieces with Walker, the notion of 'creating space' is introduced as a significant metaphor of critical conduct. Territory operates almost as synonymous with sovereignty and it signifies all claims to fixed meaning (of identity, time and space). Hence, 'to territorialize' is a metaphor for sovereignty (imposing meaning with a fixed centre), whereas 'to deterritorialize' acts as a metaphor for decentring. Meanwhile, territory seems to signify anything from 'territories of theory,' which are affirmed at the supposedly sovereign centres of the discipline (Ashley and Walker 1990: 263), to the territory of the state, which, in a similar fashion, fixes the space, time and identity of political life. In effect, space is treated as any sort of textual room in which it is possible to give a voice to any sort of identity such as class, identity, race, and marginality. However, there is also another meaning emerging; along with identity and time, space appears to characterize the human condition. And, if we accept that there is a world outside text, then there is a potential tension here between 'space in texts' and 'text in space'. By this, I mean that Ashley and Walker look at space as it appears in texts, whereas they refrain from paying much attention to the fact that texts exist in space. If we accept this distinction, it implies a difference between the territory of theory as a textual one, and the territory of the state as an (also) inhabited one.

In contrast to this approach, Henri Lefebvre, whom Walker explicitly cites (1993: 128), argues that there is a tension between the mental space in and of language on the one hand, and between social – or lived – space on the other (Lefebvre 1991: 5). Space in itself as an abstract category tells us nothing about its social or political 'content'. Therefore, it seems somewhat peculiar that, despite encouraging us to ask how spaces have come into being, that is to look at the practices, violence and power relations that constitute a space, Ashley (1987) limits his own contribution to a metaphorical abstract employment of space. In a general commentary on postmodern writers, Massey is puzzled 'by the lack of explicit attention they give [to space], irritated by their assumptions, confused by a kind of double usage (where space is both a great 'out there' and the term of choice for characterisations of representations, or of ideological closure)' (2005: 18). While on the one hand, (modern) space is associated with fixity of meaning, it is, at the same time, rejected for being untenable in a fluctuating world. Yet, even though these arguments are accompanied with encouragements to investigate the actual production of different spatial constellations, the arguments themselves are articulated in such an abstract and metaphorical manner that the meaning of space is blurred into an emptiness of meaning. It seems to signify anything from a metaphorical description of a room in a textual universe, in which one can think 'non-mainstream', to a universal condition of human existence. Furthermore, because of this 'imprisonment' of space in a cage of abstraction, it is virtually impossible to establish what the connection is between space as a human condition and the metaphorical space of IR. This is why I suggest that poststructuralism, 'Ashley style', has been better at spacing knowledge – that is to introduce new, and challenge, important spatial metaphors into our understanding of knowledge – than to provide knowledge about space.

An important contribution of the poststructuralist writers has been to illustrate how dominant theories of politics are building on specific spatial imaginations which play into claims to the sovereign political identity of modern politics. In that light, they have been profoundly successful in questioning what we think we know about politics and space and they have contributed to historicizing this knowledge. Yet, in doing this, they induce a problematic distinction between a static modern geography and a postmodern politics in spatial flux. In the attempt to rid space of a fixity of meaning, it is overridden by time, and thus reproduces Marx's problematic dictum. As Massey argues, space cannot be annihilated by time (2005: 90–8).

Bringing space back in

As the questions concerning the political organization of space and the international order have been raised there has been a telling rapprochement between Political Geography and IR (even if geographers discuss IR more than IR discusses geography). While the traditions of Political Geography and IR share a common heritage in the geopolitical tradition they developed in different directions in the post war world. Partly in response to the negative connotations of geopolitics, geography made attempts to depoliticize the study of space and politics where, for example, the biological metaphors from the geopolitical tradition was replaced by metaphors from physics (Delaney 2005: 41; see also Sack 1980). Put bluntly, then, space largely disappeared from mainstream IR theory, while politics disappeared from mainstream Geography. As space was taken as a given universal concept, it was not the concept itself that attracted attention, and the relationship between the state, sovereignty, and territorial space was not a main question. The reinvigoration of the concern with space proper, and the relationship between space and politics, was to a large extent driven by a resistance to rational planning and a scientific colonization of the concept, as I will discuss in the following chapter. Another, but related, main impetus came out of critical investigations of the global distribution of wealth and global patterns of production.

However, as has been argued by Felix Driver, the early contribution from a Global Political Economy perspective generally suffered from an overtly reliance on the economy, reducing explanation to a single function of the economy and conceptualizing the state as simply a function of the economy (1991: 269–72). That is, territorial space in these accounts remains ultimately a function of the economy. In more recent contributions to political economy, however, these claims have been strongly modified, and as pointed out by Neil Brenner a decade ago, the radical opposition between state and a global market has become more nuanced with the focus on the changes that the state is undergoing (1999). Indeed, Saskia Sassen – following the world system tradition – suggests that 'while globalization leaves national territory basically unaltered, it is having pronounced effects on the exclusive territoriality of the national state – that is, its effects are not on the boundaries of national territory as such but on the institutional encasements of that national territory' (2000: 372). Relational theories have introduced network thinking asserting how spaces are construed through a network of social relations (sometimes the social include nature as well, as we will come back to in the following chapter) promoting a non-territorial understanding of space and politics (Jones 2007: 13), and several authors have all contributed to integrate post-Marxist, or poststructuralist, insights into the study of a geographically informed political economy (Sassen 2006; Agnew and Corbridge 1995; Brenner 2004; Lash and Urry 1994; Palan 2003; and Cameron and Palan 2004). While these approaches add much to the understanding of the possible reorganization of political space, they still prioritize the international economy – its practices and organization – as the most important force in the analysis of the political organization of space, and in that respect provide a risk that space and politics become epiphenomena of the economic processes in question.

The second impulse against the 'scientific entrapment' of space has led to a concern with the relationship between power and space inspired by poststructuralist writings similar to those discussed above. The term Critical geopolitics was coined in the late 1980s and is particularly associated with the works of Gearóid Ó Tuathail, Simon Dalby (Ó Tuathail 1996; Dalby and Ó Tuathail 1998), and also John Agnew and Stuart Corbridge (Agnew 1998; Agnew and Corbridge 1995). The term *Mastering Space* represents 'a metaphor for the character of the international political economy of the past 200 years. It refers to the intrinsically geographical processes of disciplining, subjugating, exploiting and developing places that have gone on in different ways during this period' (Agnew and Corbridge 1995: xi). While Agnew and Corbridge draw on Henri Lefebvre's triadic conceptualization of space, which involves practices, representations, and daily life expressions of space, and emphasize the significance of discourse for the study of space, there still tend to be an economic bias in the analysis, like the approaches mentioned above. In his book *Critical geopolitics*, Ó Tuathail deconstructs the concept 'geopolitics' showing its 'con-*textual* nature'. This involves a reading of geography as a way of disciplining space. Following poststructuralism, this power is said to rest on Cartesian assumptions (subject/object) and a Euclidian notion of space as being infinite and homogeneous. He reads geopolitics in a context of consolidating nation-states (at home) and their efforts to expand empires (abroad). In terms of method, he works with four assumptions inspired by the writings of Derrida and Foucault: first, he underlines the importance of the power-knowledge relationship; second, he asserts that the power of modern geography is located in the centralized state; third, he investigates institutionalized ways of seeing and displaying space; and finally, he aims at resistance to the writ of the imperial state. Generally, he brands geopolitics as 'the politics of writing global space' (Ó Tuathail 1996: 18).

Being inspired by Ashley and Walker, Ó Thuatail is also critical of them for remaining on too abstract a level in their discussion of geo-power (Ó Tuathail 1996: 170–8). Despite this, his approach relies mostly on a historically informed discourse analysis. In a sympathetic critique, Nigel Thrift states that the reading of texts 'producing the world as discursive construction in a way which has problematic consequences for understanding *how* (and therefore why) geopower is actually practiced' (2000: 380). He calls for greater attention to specific material practices such as census taking, map-making and so on (Thrift 2000: 382–3). Such an attention to a broader set of practices would also problematize the extent to which power of knowledge is necessarily embodied and captured by the state. In order to analyse the relationship of the political organization of space, historically, we should not assume – based on a poststructuralist ethics – that 'the state's way of seeing' space (Scott 1998) necessarily captures the production of space. This highlights the problem of criticism or resistance, namely that it is assumed as a prior condition that there is something to criticize (often there is but this is only to highlight a potential problem with analytical assumptions carried by an ethos of criticism) thus black-boxing the critical relationship; in this case state-space.

In short, despite the increased focus on the political organization of space, a key problem that remains is exactly the failure to historicize space itself. This is to a large extent a question of what it means to make space social. While I will suggest a social study of spatial history that seeks to ascribe a measure of autonomy to space, as I will present in the following chapters, generally, in the existing literature, space is either written into an existing narrative of historical sociology broadly understood as either being driven by a state, or an economic, logic. Or space is narrowed down to a textual phenomenon that can be grasped through discourse analysis and meaning creation. And while Foucault famously admitted that he should have been more concerned with geography than he had (Foucault and Gordon 1980: 63–77), and the spatial thematics in Foucault's work is being reinterpreted (Elden and Crampton 2007; Foucault et al. 2007; Pickles 2004) there is still a sense in which the concept of space is left behind and the focus is, instead, on the social practices that subjugate space and configure it into a specific power constellation. Generally, then, most social constructivism assumes space to be a united category prior to the social practices or discourses that mould this space into various configurations,[3] and as such, they maintain a subject-object distinction placing space on the social side of this divide, and by implication, leaving the 'thingness', or 'objectivity' of space on the other side.

This subject/object divide affects the way in which space is understood in the discussion of change from a modern to a postmodern world. Globalization theory, at large, either claims that space has been compressed, or is being annihilated, by time; or, that it is a different spatial logic – that of networks – that overtake the territoriality of the modern age. Both claims are based on an assumption of space as a physical presence which can be overcome by technological innovation, and capital, speeding up social processes and removing spatial barriers to communication in real time. This renders space a physical static presence which will be increasingly less significant as humanity constantly invents more efficient ways of communicating and transporting goods and services. Now, even though it is true that the steam engine, the telegraph, aeroplane technology and so forth have had immense implications for social practice in reducing so-called spatial barriers, it is not obvious to apply this spatial logic to territory. It was not capitalism and technology by themselves that produced a territorial geography in the first place. If we reduce the political organization of space to the outcome of technological development and the needs and causes of capitalism, then we empty the territorial space of political content and the significance of geopolitical competition. Furthermore, the idea of a tension between globalization and the sovereign state also tends to naturalize the global space as a natural unified space. When this relationship is articulated as a tension between a political fragmentation into territorial spaces on the one hand, and a global market, civil society, humankind on the other, it ignores the fact that the notion of a unified global space, providing a level playing ground for all actors, is itself a political fabrication which has been closely entangled with the formation of territorial states. In other words, it is not only territorial space which is political, global space is too, and in consequence, global and territorial space, respectively, represent two sides of the same coin. Historically they emerged together and, it was against the unification of global space, that the territorial states began to take shape in Europe.

In sum, it is clear that, especially since the end of the cold war, the spatial identity of the state and, with that, the political organization of space, has become a more open-ended question undermining the static image of space that has characterized the rationalist approaches to the discipline. Peter Taylor's argument that '[t]here is nothing new about cross-cutting spaces of territories by spaces of flows [...] but the fact that it can be viewed this way [...] indicates just how embedded states have been in how we view the world' (2000: 378) illustrates well how a new focus on spatiality has contributed in significant ways both to the understanding of so-called modern politics and also to the issues that push some politics and some social practices beyond the spatial configuration of the nation state. Yet, it is also clear that these theoretical openings generally struggle to tackle the concept of space. Most accounts rely on a problematic historical distinction where a modern era, in which political space was organized in a stable territorial geography, is juxtaposed with a postmodern era in which social practices are temporally accelerated and where geographical distance matters less. Importantly, such approaches ignore the extent to which the political organization of space is not only about states, it is also about the globe, and therefore, it is essentially problematic to posit a diametrical opposition between the two in this context.

There are possibly two reasons for this: one concerns space and, the other, the state and its transition. In terms of re-invigorating the discussion about the state-space relationship, the material causality which governed early geopolitics had to be abandoned as things 'do not have meaning' by themselves. As a consequence, though, geographical space becomes difficult to tackle because it stands in between what is usually considered 'a physical reality' and 'how we make sense of it'. The consequence of the 'divide' between those that rely on a notion of 'physical nature' where territory is considered a constant physical presence, and those that rely on 'social production' and 'meaning-giving' practices is, in effect, a reification of a subject/object divide concerning space. This divide arises from a disagreement of the status of the notion of science, the purpose of social theory, and how to relate to 'reality'. One of the main tasks in the subsequent chapter is to cope with this dispute over science, which has been prevalent in the discipline over the last decade; seek to avoid the subject/object divide; and argue, in a Latourian fashion, that space is indeed a physical reality, but only because it is a social construction. It is, then, the ambition of this book to address the question of the political organization of space without maintaining the subject-object divide and without ascribing different spatial logics to different time periods. On the contrary, the aim is to establish a framework within which it is possible to articulate space as having a defining and conditioning impact on politics without rendering it an object or a 'non-material' social construction. That is, to ascribe a degree of autonomy and causality to space while maintaining it as a historically contingent category. I will approach the issue of space through cartographic history because, through cartography, space can be studied as a social construction without making it derivative of, say capitalism, ideological hegemony or state formation, but rather, as something that is being established on its own and hereby invokes a certain causality upon these other practices, which are usually considered 'makers of space'.

3
Reclaiming a Spatial Reality

I would suggest that it is only when the spatial order has *appeared* to be relatively stable, that is during the cold war years, that the political organization of space can seem insignificant. As soon as the sovereign territorial order is challenged, then the significance of space is brought to the fore in social scientific enquiry and expressed in attempts to examine what the international is *inter* between and how this is ordered. Assuming that space is both a universal condition of social practice, and as such plays a conditioning role, while at the same time space is a result of social production or construction, and as such plays a constitutive and a supportive role, the challenge that remains is how to theorize space in a fashion that avoids the spatial determinism of the geopolitical tradition without rendering space an overtly social phenomenon. In order to play a conditioning role for politics, space must be rendered autonomous, to a degree, of the social processes that they are supposedly framing. If not, space will become tied up in the greater narratives and rendered an outcome of, typically, capitalism or state formation. In other words, the task is to render space autonomous without making it a natural object.

Although Historical Sociological accounts on state formation have contributed to a richer understanding of the state, it has been less than forthcoming when it comes to the knowledge categories in which we think. Thus, the challenge posed by philosophical questions about knowledge, and the relationship between knowledge and power, have been less profound. In consequence, this tradition generally maintains a solid distinction between what is social and what is natural space. Instead space is largely considered a pre-existing and static 'thing' providing the stage for historical actors. Thus, so far I have demonstrated that both historical accounts of state formation as well as contemporary interrogations of the political organization of space, generally maintain a solid subject/object distinction in their understanding of space as well as troubled historiographies ascribing different spatial logics to different eras. In response, I will introduce Bruno Latour's science studies in order to provide the ground for a historical sociology of space. In this, space will be conceptualized as both 'social' and 'natural', and further, as something that achieves the status of an autonomous phenomenon by virtue of being something constructed. This notion allows me to historicize space as well as the state, which is a necessary move in order to counter the state-space problem outlined in the previous chapter. By making space autonomous, yet constructed, space both conditions possible forms of political organization, such as the state, while, at the same time, the state can use space as a resource for its own goals. In other words, space makes states and states also make – and use – space. Being sensitive to this duality, we can investigate how the transformation of both the state and space in a specific conjuncture provide the basis of the prevalent image of the international system.

The problem, in the words of Massey, is that the 'definition of space was caught up in the broader dialogue between the "natural" and "human' sciences"' (2005: 31). Science wars, that is, metatheoretical disputes concerning the role and possibility of science,[1] have characterized the social sciences generally, and in particular, they have contributed to the predominance of the subject/object distinction underlying the space problematique defined in the previous chapter. The discipline of International Relations (IR) was no less hit by the science wars than other disciplines; what Ole Wæver identified as a rationalist neo-neo synthesis between Neorealism and Neoinstitutionalism was challenged by a mixed group of contenders questioning the 'scientific' paradigm defended by the orthodox rationalists (1996: 156–70). The content of the contender camp varied according to who defined them and when. For example, Robert Keohane articulated the division between rationalists and reflectivists by including in the latter position postmodern, linguistic, hermeneutic and historical materialist approaches, *in fine*, anything that was not rationalism (1995: 284; Wæver 1996: 164). Ten years later, in a special issue of *International Organization*, the description was narrowed down to a certain notion of constructivism understood as clearly distinguished from postmodernism, which was perceived as operating beyond the enterprise of social science (Katzenstein et al. 1998: 677–8). Within Geography, much of the impetus for Human Geographers, as I will discuss subsequently, exactly came as a response to an overtly simplistic and science-based understanding of space. And as Kirsten Simonsen states, in her discussion of 'the problem of integrating the material environment into social theory', there is a fundamental distinction between the natural and the social sciences (1996: 497).

There have been plenty of attempts to bridge the great divides informing these conflicts, but only few have sought to transcend them. Alexander Wendt's predominant version of social constructivism presented in *Social Theory of International Politics* (1999), and further elaborated in a more recent essay (Wendt 2006), is built up around this fundamental distinction between the mind and external reality, and his wish to bridge this gap. In his own words, Wendt's enterprise seeks to provide a *via media* between his rationalist scientific epistemology and an 'idealist ontology' (1999: 40). This, indeed, reaffirms the subject-object distinction and, in consequence, he is not able to provide more than a stable truce. Giving up the search for the 'one true description', Wendt's solution to the science wars is 'an 'epistemological Westphalia', in which positivists and interpretivists recognize the other's contribution to their shared goal of comprehending social life' (2006: 216). Bringing a notorious treaty to end the conflict, the culture/nature distinction stands firm, as does the problem of either locating the reality of space inside the mind, or out there, in the world.

In response to the bridge-building between a natural and a social world, as well as demarcation exercises, where the social is fended off as a special realm, Bruno Latour states that it is 'the battle of the science warriors and that alone that forces us to believe in opposing camps, to rally and align ourselves as if there were a battle' but, as he continues, 'there isn't a battle' (2002a: 76). He emphatically rejects the notion of two opposing camps harbouring an objective science and a subjective constructivism respectively. Instead he suggests that what is typically considered 'nature' in the social construction of space should be included as an active agent in 'social construction', that is, to give agency to something other than humans. Instead of starting off from the usual distinctions between subject and object, and culture and nature, the aim is to investigate how knowledge technologies of space serve to establish a specific reality of space; it becomes real by virtue of being constructed, and it is constructed as an assemblage between human and non-human elements. In that respect space is generated trough particular networks that tie together the landscape with models of calculation that all, in sum, establish space as a real and autonomous phenomenon.

Knowing real space

There is an intricate relationship between knowledge and space which makes the issue of knowing 'real space' more complicated than most questions. Spatial metaphors are often associated with knowledge production; not least in terms of mapping an area of knowledge, and on a more essential level, at least since Kant, space has played a central role for the possibility of knowledge. Discussions therefore easily get philosophical to the extent that they lose sight of empirical analyses or they remain unsatisfying, seen from the philosopher's perspective. Yet, running the danger of falling into these traps, I believe with Latour, that a degree of so-called metatheoretical discussion is unavoidable:

> [C]ase studies, philosophers argue, do not concern themselves with the 'foundations' of science or with the 'transcendental conditions' of any argumentation. There is thus a division of labor between philosophers of science, who think they have a perfect right to ignore (and even to despise) empirical studies, and the social scientists, who think they should never indulge in philosophical arguments. [...]. This division of labor is a catastrophe; philosophy and field studies should be carried out under the same roof and, if possible, in the same head.
>
> (Latour 1988: 252, note 7)

While, in philosophy, there is a strong movement to transgress a Kantian framework of science that relies heavily on a distinction between the mind and the body, between an inside and an outside of the subject, Latour turns the practice of science into his object of study while also examining its philosophical underpinnings. This illustrates the contingency of 'reality' or 'objective knowledge' on practical grounds and not only through a deconstruction of its metaphysical underpinnings. And this contains the key to escape the problems of the subject/object that has prevailed so profoundly in the discussions of the political organization of space.

In *The Critique of Pure Reason*, Kant posits space – along with time – as a transcendental category inherent in the subject as a condition for being able to know anything about reality. At the same time space and time are not only 'transcendentally ideal' (pure forms of intuition, *Anschauung*) they are also 'empirically real' (Guyer and Wood 1998: 7). Space is both something that can be experienced empirically and a pre-condition for experiencing anything at all (Madsen 1994: 18–19).

The problem, according to Kant, is that it is impossible for the individual being to grasp the essence of objects. Objects can only be known through our experience of them – as they appear to us. This statement renders knowledge subjective as it is only through experience that we can know anything about the real world. Descartes' theory of knowledge divides things into *res cogitans* and *res extensa*: between consciousness (without extension) of the subject, and matter with an extension in space and time (Madsen 1994: 18; Lübcke 1990: 369). In Kant's philosophy, space is not primarily a substance in itself but formalized through geometry entrenched in the subject as a 'form of intuition' (*Formen der Anschauung*), which becomes a condition for knowing reality (Madsen 1994: 17–18). Yet, this does not entail a total 'subjectivisation' or a 'relativisation' of space. The categories of space and time can only make a universal science possible if they are reflected in the order of the real world outside the subject. Hence, universality is reintroduced through the categories of time and space exactly because these transcendental, or *a priori*, forms of intuition do not depend on experience. They are stable and make universal experience possible.

We therefore assert the empirical reality of space (with respect to all possible outer experience), though to be sure its transcendental ideality, i.e., that it is nothing as soon as we leave aside the condition of the possibility of all experience and take it as something that grounds the things in themselves.

(Kant 1998: 177)

Thus, the conditions of possibility of experience are the universal categories being located in the subject and it is, therefore, the subject that provides the possibility of scientific knowledge (Guyer and Wood 1998; Kant 1998: 172–84). Kant describes this as a 'Copernican Revolution', by which he compares his own move to make the world revolve around the subject (anthropocentrism) with Copernicus revolution in astronomy, by moving from geocentrism to heliocentrism (Guyer and Wood 1998: 70).

Now, it is evident how the poststructuralist dissolution of the individual as a coherent entity affects the possibility for a certainty of science as aimed for by Kant because this possibility rests on a subjectivity universally containing the transcendental categories of time and space required for such an enterprise. A decentred subject can no longer be the source of true cognition, and universal knowledge is therefore no longer possible.[2] In consequence, space is relativized and becomes a subjective feature.

However, the subjectivisation of knowledge as a result of decentring the subject is, to paraphrase Latour, to reproduce the Kantian model of a 'mind-in-the-vat'. Latour describes Kant's philosophical solution to the Cartesian search for certainty as a disaster in terms of improving our understanding of the world. The core of the problem is the firm detachment of the mind, or the inner life, from the body and the world (Latour 1999: 5–6): a mind in here (in the vat) and a reality out there. Crucially for Kant, these two spheres are governed by different principles. Put simply, time 'determines the relation of representations on our inner state' (Kant 1998: 180), whereas 'all outer appearances are in space and determined *a priori* according to the relations of space' (Kant 1998: 181). The location of knowledge (the mind) thus becomes firmly separated from the reality of which it seeks knowledge. And, logically, if the subject is no longer the carrier of universal *a priori* categories, a universal knowledge of reality is no longer possible.

In Latour's humorously caricatured reconstruction of the history of the philosophy of knowledge, Kant's 'mythical mind' giving shape to reality 'was soon replaced by a more reasonable candidate, *society*' (Latour 1999: 6). However, this substitution only replaced the subject with 'the social group', but still referring to an external – and thus separated from – reality.[3] These reflections indicate that the main conceptual problem is to be found in the differentiation of the mind – an inner world – from an external reality, and as long as we remain within this frame we will ask the wrong questions. Latour asks us to retrace our steps back to Kant and question what he labels the 'modernist settlement' which involved the differentiation of 'the mind' from 'nature' (epistemology), and 'society' from 'god' (politics and morality). In opposition, he maintains that '[w]e do not need a social world to break the back of objective reality, nor an objective reality to silence the mob' (Latour 1999: 15). By this he means that we do not have to retain a notion of objective reality in order to avoid an anarchic relativisation that is often put forward as a counterargument to social constructivism.

It could be argued that Latour sets up a straw man that is too comprehensive to be convincing. Establishing all prior distinctions between science and politics only to demolish them with a single blow is not persuasive. Nevertheless, his 'science studies' and arguments concerning the production of knowledge contains a promising approach to space; while his solution to the science wars is that 'we are *not* at war' (Latour 1999: 15), he is advocating a 'realistic realism' which maintains a notion of reality – not as something to believe in but something which is absurd to leave out because we will never stop evaluating knowledge claims. At the same time, he maintains that facts are fabrications but this is based on specific studies of how facts are established. In that way we can trace how a specific spatial reality is established without rendering it something that takes place beyond the mind, in nature, as being static and without meaning.

Such a solution is needed in order to depart from the Kantian framing of space as something universal yet something residing in the subject, governing our experience of the 'outside' world, which has done much to enforce the reliance on, and choice between, *either* physical or social space. As noted, Massey suggests that space has been caught up in the science wars, where 'social space' is deployed to attack or provide an alternative to an exaggerated reliance on a 'scientific', thus fixed and natural, understanding of space (Massey 2005: 20–35). Because of this, the attributes of a 'fixed meaning' and 'Euclidian' characteristics has been associated with 'nature' and consequently subject to critique or reification. Hereby, it is ignored that space can be said to possess such characteristics – not due to its inherent nature, however, but rather as a consequence of the way in which space has been established through social practices.

As discussed in the previous chapter, space was largely put back on the research agenda in order to emancipate the concept from an exaggerated reliance on science. One of the main figures in bringing space back on a critical social studies agenda is Henri Lefebvre.[4] A key target for his critique is the (Kantian) belief that the reality of space corresponds with the abstract measures of Euclidian geometry. He famously opens his seminal *The Production of Space* by stating that

> [n]ot so many years ago, the word 'space' had a strictly geometrical meaning: the idea it evoked was simply that of an empty area. In scholarly use, it was generally accompanied by some such epithet as 'Euclidian', 'isotropic', or 'infinite', and the general feeling was that the concept of space was ultimately a mathematical one.
>
> (Lefebvre 1991: 1)

It was very much a critique of a scientific representation of space which provided the drive of writers like Lefebvre, Bachelard and de Certeau's (1988) concern with space. These writings were concerned with bringing 'to life' those elements of space that were not captured by the planning and official practices of the modern state (in Europe). These writers attempted to prioritize 'the lived space' over that of space planned and calculated. In the words of Bachelard: 'In [the] dynamic rivalry between house and universe, we are far removed from any reference to simple geometrical forms. A house that has been experienced is not an inert box. Inhabited space transcends geometrical space' (Bachelard 1994: 47). This quote nicely illustrates the opposition to a Kantian position wherein geometrical space transcends the experience of space as the condition of possibility for universal knowledge.

These writings were therefore attempting to give meaning to the history of space as something produced and lived rather than as something abstract and pre-existing. In attacking the contemporary emphasis on space as a concept, Lefebvre states that knowledge ought not to make representation the basis of social life:

> The object of knowledge is, precisely, the fragmented and uncertain connection between elaborated representations of space on the one hand and representational spaces (along with their underpinnings) on the other; and this 'object' implies (and explains) a subject – that subject in whom lived, perceived and conceived (known) came together within a spatial practice.
>
> (Lefebvre 1991: 230)

The aim here is to emphasize how the reality of space is something that has been produced by social practices over time and, hence, something which is lived and alive rather than an empty and static substance that can adequately be described in terms of a three-dimensional absolute space. This spatial criticism was combined with a resistance to rationalization and state planning of the environment in late capitalism. 'To change life', states Lefebvre, 'we must first change space' (1991: 60; see also Dear 1997).

Authors like these have been instrumental in generating a significant concern with space in Sociology and Human Geography wherein space was established as an inescapable dimension of all social life (Simonsen 1996: 505; Crang and Thrift 2000: 3; Soja 1989). As space was no longer seen as something given, or absolute, a wide array of studies into the relationship between social practice and spatiality opened up.[5] With the focus on production, Lefebvre implied that space is produced differently in different societies, hence, space becomes relative to its mode of production. In critical socio-spatial theory, the link between space and capitalist modes of production has been a common trait. For example, David Harvey's *The Condition of Postmodernity* argues that the objective qualities of physical space cannot be understood independently of material processes. It is the material practices of social reproduction that gives objectivity to space: 'Each distinctive mode of production or social formation will, in short, embody a distinctive bundle of [...] space practices and concepts' (Harvey 1989: 204). This, in effect, ties space very closely to the concept of time, as space is produced differently in different historical contexts. Elsewhere, Harvey points to the anthropological evidence that different societies produce qualitatively different conceptions of space and time and, further, that a 'particular way of representing space and time guides spatial and temporal practices which in turn secure the social order' (1996: 443). A sensitivity towards the way in which space is produced in different social orders breaks away from the Kantian notion that the reality of space is absolute.

However, as discussed in the previous chapter, there is a danger to overemphasize the 'social' of space in a particular way that makes space a product of capitalism. In Harvey's text, space ultimately appears as a derivative of the mode of production, even though 'the state' is mentioned, the command over space is seen as a crucial element in the search for profit and, indeed, Harvey specifically ties control over space to class struggle (1989: 232). Thus, whereas Harvey successfully remedies the absence of space from historical materialism at the time (Gregory 1989: 186–7), he remains within a meta-narrative of space being propelled by 'modes of reproduction'. And a reliance on 'mode-of-production' thinking is not an effective tool to understand the historical processes through which meaning has been given to the spatial identity of the state. If space is not granted a measure of 'autonomy', it will unavoidably end up being a result of the processes that are being explained. Because Harvey's argument is unable to break out of the mode of production as the cause of spatial transformation, it would essentially suggest that the mode of production determines possible modes of political organization. Yet it is vital not to equate this 'analytical autonomy' of space with a reliance on nature. Rather, the challenge is to render space both 'natural' and 'social' instead of black-boxing 'nature' and only focusing on the 'social'. In her account of space, Simonsen argues that 'the material environment cannot be part of a social theory except through the mediation of social practices' (1996: 506), and hence, the way to relate to the material environment is to address the social practices in relation to this. Such an approach inevitably reinforces the nature/culture divide by considering the 'natural environment' as something which is simply there independently of social agency and history. Nature is not an absolute and stable space, and should not necessarily be left to the natural scientists.

Re-establishing a reality of space

Inspired by Latour, Nigel Thrift has suggested that space should be considered as 'the outcome of a series of highly problematic temporary settlements that divide and connect things up into different kinds of collectives which are slowly provided with the means which make them durable and sustainable' (Thrift 2003: 95). While Latour shows clear poststructuralist influences, he presents a different approach that defines itself in opposition to both modernist and postmodernist understandings of knowledge production. Drawing on ethnomethodology and pragmatist philosophy, his work is characterized by intuitive insight combined with very complex models and descriptions of how knowledge is *fabricated*. He is famous for studying the processes of science as well as questioning many 'usual suspect assumptions' about *reality*. The main aspect of this argument is that the commonly upheld distinction between 'language' (linked to the subject) and the 'world' (as an object) is flawed. The implications of an 'ontological detachment' of language from the world is that statements (of knowledge) will have to relate to the real world out there, or we can say that there is no access to this real world and hence we focus on the world that is constructed. Either the evaluation of knowledge claims is based on a comparison with a foundational reality or there is no foundation for evaluation. Admittedly, this is a caricatured depiction: constructivist and discursive approaches do not advocate total relativity. Nevertheless, the point made by Latour is that as long as we maintain an essential gap between world and language the problem will essentially remain.

The solution he offers is, firstly, to argue that there is no essential gap – world and language are not separate domains. Instead of operating with a notion of referents that either represent or perform a world – and thus bridge the gap between world and language – he adopts a notion of *circulating referents* that circulate between *humans* and *nonhumans* in a continuous process. This is a way of re-conceptualizing the Kantian subject-object divide. Simply speaking, the point is that both humans and nonhumans affect the referent and hereby propositions are not a property of the human only. In short, *nonhumans* 'have agency too' and also decide propositions. I read this as a way of 'giving causality' to 'the material', however, without in any way pre-defining this causality. Even more importantly, *nonhumans* is not a constant – nature is not a constant. This is not an argument about how forests grow, mountains erode or volcanoes erupt, but rather it is an argument saying that the world of *nonhumans* enters a relationship with *humans*. They form what Latour calls a collective that both transforms *humans* as well as *nonhumans*. As a phenomenon, space does not occur in the sudden encounter between space as 'a thing in itself' and categories of human understanding as discussed above (Latour 1999: 71). Instead, Latour argues that 'phenomena are what *circulates* all along the reversible chain of transformations, at each step losing some properties to gain others that render them compatible with already established centers of calculation' (Latour 1999: 71–2). The 'reversible chain of transformations' describes the process through which scientific facts are established. This is a process of several stages in which observations stage by stage get rid of 'locality, particularity, materiality, multiplicity, and continuity' (Latour 1999: 70) – in a process called *reduction* – where what is observed is plotted down on a piece of paper or a computer. Simultaneously, at each stage of reduction the representation obtains greater 'compability, standardisation, text, calculation, and relative universality' (Latour 1999: 70). Hence a phenomenon circulates in-between particular observations and a generally 'universal' body of knowledge with which it is made compatible in its transformation to a fact. This is a somewhat complicated argument, yet phrased simply, phenomena are not something that either materially exist or achieve formal existence only through language, but rather, are something that achieve an identity through many stages of interaction between *humans* observing, testing and interacting with *nonhumans*.

A further blow to the subject-object divide arises from Latour's discussion of 'collectives of *humans* and *nonhumans*' where he posits 'society' as a collective not only of *humans* but also of *nonhumans* (Latour 1999: 174–215). By the phrase 'objects have agency too' Latour (2005: 63–86) argues that objects and technology, for example, alter the motivation and the goals of 'actors' when they enter into a relationship. Using the example of the dispute concerning gun control in the United States, he argues that neither the statement *guns kill people* nor the counter *guns don't kill people; people kill people* provides the correct answer (Latour 1999: 176–7). The former lets technology decide and the later reifies the agency of 'people'. Instead the argument is that when the human and the gun enter into a relation they are both transformed. In a sense it is a trivial point, but the wider implication is that objects ought to be included in what is considered social and, furthermore, that society cannot be explained without giving proper agency to all the techniques and 'objects' which are essential for its constitution.

In accordance with other critics of the notion of 'absolute space', Latour argues that

> [m]ost of the difficulties we have in understanding science and technology proceeds from our belief that space and time exist independently as an unshakeable frame of reference *inside which* events and place would occur. This belief makes it impossible to understand how different spaces and different times may be produced *inside the networks* built to mobilise, cumulate and recombine the world.
>
> (1987: 228).

Opposed to the Kantian notion of space and time residing as static and universal categories in the subject, space and time, for Latour, is something that is being fabricated within networks and what he calls *centres of calculations*. Rather than being a natural frame, 'space is something generated *inside* the observatory' (Latour 1987: 229). That is to say that space, as a phenomenon, achieves reality as an autonomous thing at the specific site in which measurements and observations are assembled and represented. Hence, even though 'Western Science' claims to be universal, it is by its very nature a very local achievement. Even a universal cartographic representation of the earth or the universe is put together at a specific site, in a particular workshop.

The fabricated notion of space is not trivialized by Latour; rather, it seems that space is a requirement for other scientific practices. He hints at this in *Pandora's Hope*, when he comments on the significance of two maps used on a scientific expedition in Brazil:

> Remove both maps, confuse cartographic conventions, erase the tens of thousands of hours invested in Radambrasil's atlas, interfere with the radar of planes, and our four scientists would be lost in the landscape and obliged once more to begin all the work of exploration, reference marking, triangulation, and squaring performed by their hundreds of predecessors. Yes, scientists master the world, but only if the world comes to them in the form of two-dimensional, superposable, combinable inscriptions.
>
> (Latour 1999: 28).

This is so because science, or more generally action, cannot take place in a universal and abstract space. Science practice needs to be able to navigate, collect, assemble and combine with existing knowledge in other sites. This argument asserts the centrality of knowledge of space: in order to do something planned somewhere, this somewhere has to be known. Otherwise one would enter as a stranger. Imagine going to an unknown city without a map; the only way to find somewhere would be to ask the locals for directions – to rely on local knowledge. If in possession of a map, the knowledge of the locals is irrelevant for you and you would be easily able to navigate the city and get to your destination on your own.

Hence, the coordination of social practice *in* space requires that space is made present, known and navigable through a technology or mode of representation that translates the landscape and environment into a coherent notion of space. This is, in other words, to give primacy to cartographic practices for understanding the reality of space and by implication, that histories of cartography represent histories of spaces (see Pickles 2004 for a related argument). Space as a phenomenon is established through cartography as, what Latour calls, a *circulating referent*. Space, as a term, does not signify any specific object in the way the word 'tree' signifies a specific organism. Space both points to distances and locations but also to specific *nonhuman* features such as land. What is significant is how these features are being included into the establishment of space. Referring to the example about navigating a city, knowledge of space is about familiarity (Latour 1987: 220). Hence, people will always have knowledge of the space with which they are directly familiar. However, action at distance (larger-scale coordinated action such as trade or scientific expeditions) requires familiarity with the places one seeks to act at. In order to do this, unknown sites have to be made mobile and compatible with what is already known. Compatibility is almost self-evident because if you wish to add to what you know, new knowledge has to have a form which makes it possible to combine with what you already know. Mobility is maybe less self-evident, but not less important. For an unknown place to be made familiar, knowledge of this place has to be combined with knowledge at another site (since all knowledge production is local). Hence a system of referents has to be employed in order to make sites *mobile, keep them stable* and *combinable*, in order to combine various places. Latour names the place where such referents are combined a *centre of calculation* (1987: 223). The centre of calculation is a site, a laboratory for example, where circulating referents are combined, calculated and developed. In the context of space, it is the site where places are made mobile, assembled and established as a reality of space.

Scientific cartography does all the above, it allows places to be rendered mobile, kept constant and combinable (Latour 1987: 223). Hence, the map-makers' workshop, or a royal court that controls such a process, will act as a centre of calculation. They constitute a hub in a network of navigators, surveyors, cartographers, printers, etc. which are all necessary for geometric cartography to develop. 'Cartography is one network cumulating traces in a few centres which by themselves are [...] local' (Latour 1987: 229). But cartography also allows this centre to act at a distance on other places that have become available to those at the centre. To give examples, a king would be able to capture and plan travels in the entire kingdom by looking at a sheet of paper on his wall; the Dutch East India Company would be able to plan the route of fleets in their map room and so forth.

Rather than reaching new land, then, the significant point of the early European expeditions to the rest of the world was to bring the new places back home in order to facilitate a return voyage. Without this ability, the trip would be wasted; the ships would disappear in the horizon and not participate in enlarging known space. 'By coding every sighting of any land in longitude and latitude [...] and by sending this code back, the shape of the sighted lands may be redrawn by those who have not sighted them' (Latour 1987: 224). Thus, in the words of Latour, 'the cartographers in Europe start gathering in their chart rooms – the most important and costliest of all laboratories until the end of the eighteenth century – the bearing of all lands' (1987: 224). The consequence of this massive endeavour is a transformation in which cartographers come to dominate the world, which means that the *nonhuman* aspect of space loses power, in the sense that capes, corners, vast spaces become less and less of a danger and an obstacle to trade and travel.

Departing significantly from Kant's space as a universal category, Latour provides an alternative that allows us to talk about, and study, a reality of space. Significantly dissolving the gap between language and world, the point is no longer to argue whether the world (material) dominates or whether language (ideas) dominates, but instead to trace how space is fabricated at particular sites. Emphasizing the specific locality of the fabrication of space, Latour's framework allows and encourages us to trace more specifically how knowledge is produced by opening the 'black box' of science. By focusing more specifically on the agency and the techniques involved in cartography, primarily focusing on geometric cartography, we can achieve a much more nuanced view of the spatial identity of the state and how the cartographically established territory has been giving meaning to the space of international relations.

Historical sociology of state space

Whereas Historical Sociological accounts of state formation remains silent on the historical transformation of space, Latour's version of actor-network theory is virtually silent on the concept of the state. He does not aim to trace the assemblages that make up the state as a collective, as he would likely describe it.[6] Inspired by Latour's understanding of collectives and associations, Nigel Thrift has described the nation state as an 'actor network'. Nevertheless, in order to deal with questions concerning the spatial identity of the state, and state formation, we have to include a discussion of the power of space as well as a discussion concerning motivation in the sense of why the state could or would use space in order to pursue particular goals or strategies. And while Latour is busy acclaiming the novelty of his sociology, Historical Sociology and Latour's actor-network theory share similar purposes. Michael Mann's project, for example, to trace how different overlapping socio-spatial networks of power have combined historically in different socio-political figurations (Mann 1986: 1–3) resonates closely with Latour's repeated statement that sociology should be about 'the *tracing of associations*' (Latour 2005: 5). Hence, despite obvious methodological differences, Latour and Historical Sociology share enough of the same concerns that it is feasible to integrate a Latourian sociology of knowledge with a historical sociology of the state.

A well-established battle ground within the Historical Sociology literature is occupied by those that seek to explain European state formation in terms of geopolitics and interstate competition and those that emphasize class relations.[7] Charles Tilly (1992) famously highlighted the role of warfare, and in that respect he positioned himself in the vicinity of Otto Hintze who, in 1906, wrote that '[a]ll state organization was originally military organization, organization for war' (1975: 181). Arguing against Marx, Hintze suggested that class conflict was not the only driving force of history. The conflict between nations has been more important and pressure from without has been a determining influence of an internal structure (Hintze 1975: 181: 183–4). The emphasis on war and geopolitical tensions as the main drive behind state formation is frequently labelled a Neo-Weberian approach (associated with writers such Giddens (1985), Mann (1986; 1993) and Ertman (1997)). Criticizing the Neo-Weberians, Benno Teschke, drawing on 'Political Marxism', presents a strong argument centred upon social property relations as mediators between major classes that define the constitution and identity of political units (2003: 7–8). He shows how international relations 'are internally related to politically instituted class relations (social property regimes)' and how geopolitical pressure affects the course of socio-political development (Teschke 2003: 272). Geopolitics, or 'the pressure from without' as Hintze would have it, in this model is internalized and deciphered as 'social praxis' linked to prevailing social property regimes (Teschke 2003). Being critical of both Marxist accounts, for focusing on domestic dynamics while abstracting them from the geopolitical contexts, and Neo-Weberians, for overemphasizing power politics as conducted by the state without taking the social property relations into account, Teschke (2006a: 567) tries to navigate in between in his attempt to trace modern international relations. While ultimately ascribing the explanation to social property relations, he also admits that 'geopolitical plurality' and the diffusion of states cannot be explained or derived from 'capitalism'. On the contrary, 'capitalism arose historically in a multi-territorial matrix that was the legacy of the absolutist period' (Teschke 2006a: 540). And it is this territorial matrix that the historical sociology of space can help us to understand.

The disputes of causality in the state formation literature has been much concerned with the notion of the state's relative autonomy (from the mode of production), and in particular Michael Mann (1984) has been ascribing analytical autonomy to the state in order to capture the historical constitution of society as well as understanding state formation. I have already discussed how space should be granted relative autonomy in order to understand how space has conditioned processes of state formation historically. In other words, we should grant 'relative autonomy' to both space and to the state for the purpose of understanding how a cartographic transition transformed the conditions for the political organization of space. As will be shown, the transition was one in which the state tried to take part and tried to control. But not until a late stage was this control achieved by the state and the transition was thus not exclusively state driven. Therefore, the transformation of space should be seen as being autonomous, to a degree, from state formation. This is not because any sphere of activity should have 'ontological primacy', but because it illustrates how the development in state structures interrelates with space. If one were to establish a single causal analysis, based on a single set of social relations, the state and space would be derived from the same source and it would not be possible to relate them to each other the way I set out to do. To recap, the point is not to explain the causes of a cartographic transition, but instead, to show how new cartographic practices conditioned the process of state formation.

So, space conditions social practices, but it also provides a means of action and it is, therefore, a source, and a consequence, of power relations. Spatial practice 'masters and appropriates space', but also presupposes it. Lefebvre's understanding of the historical production of space has provided eloquent arguments concerning the power of space. Space reflects a history while it also designates possible future histories: 'Itself the outcome of past actions, social space is what permits fresh action to occur, while suggesting others and prohibiting yet others' (Lefebvre 1991: 73). Lefebvre argues that space is a concrete abstraction and it can thus be captured dialectically: '[t]he concrete abstraction is simultaneously a medium of social actions, because it structures them, and a product of these actions' (Shields 1999: 160, quoting Gottdiener (1985)). Space is both a condition of possibility, but also an effect of social agency and, translated into the relationship between the state and space, this becomes one of mutual conditioning. With reference to the gun example discussed earlier in this chapter, both the notions that 'space frames people' and that 'space doesn't frame people; people frame people' would be misleading. Rather, the point is how specific assemblages, or fabrications, of space conditions goals and possibilities and this is, in a sense, to give agency to both space and people. Obviously 'natural barriers', which are frequently emphasized when space is considered 'nature', such as the island figuration of Britain, have had implications for strategic considerations and Britain's ability to trade, wage war etc. Yet, it is the way in which these are assembled to establish space as real that conditions what practices are possible. An obvious example of this is how geometric map representations of the globe allowed certain navigational practices which would not have been possible under another regime of spatial representation.

However, in addition to playing a conditioning role, the reality of space also serves as a means to power. Spatial control serves as means to control other practices relating to this space, and the possible mode of controlling will evidently also be conditioned by the specific fabrication of spatial reality. Historically, territory was a source of income, directly as farmland and taxation rights, and indirectly as a means to control other practices such as trade rights. War created a constant demand for money and geopolitical power struggles led to an ever-increasing demand for more resources in the shape of finance and soldiers which further demanded more efficient administration and governance of the state territory (Gustafsson 2000). It is not only in the context of state power that control of space is a means to other things. As Harvey notes, command over space and time is a crucial element in the search for profit (1989: 226). And, by the turn of the last century, it was commonly emphasized how geography as a discipline opened space for other activities such as commerce and investment. For a pertinent example, H. M. Stanley linked geographical knowledge and cartography in his opening address to the Scottish Geographical Society in 1884: 'geographical knowledge clears the path for commercial enterprise [which is the beginning of civilization].' He notes further how mapping rationalises space: '[t]ake up any old map of Africa, and glance at the antique and grotesque creations of the Portuguese missionaries and travellers of the fifteenth and sixteenth centuries, and compare it with that of to-day, illustrated by the travels of nearly 800 explorers. It is only now that we begin to have a rational idea as to what Africa is, and whether commercial enterprise is in any way possible' (Stanley 1885: 4). There have thus been a number of interests involved in the mapping of the world, and the establishment of a global spatial reality. Yet, when scrutinizing the spatial identity of the state, I emphasize how altered practices in establishing a reality, historically, transforms space from being a vague source of income to more diffuse power relations because as space becomes a defining aspect of the state's identity more and more relations are mediated and controlled via this space. If, as Michael Mann argues, political power is about drawing boundaries, then the reality of space decides how boundaries can be drawn, policed and exploited, and therefore we must give space a history as well.

By implication of following a Latourian perspective, space understood as a phenomenon being *human* as well as *nonhuman* has a history too. To paraphrase Latour: 'we decided to grant history to space, not only to the humans discovering it' (1999: 146).[8] And again, maintaining a desire to speak in terms of 'reality', this is not sufficiently done in writing a history of the concept of space in human thought only. Rather we can historicize space as something *nonhuman* without falling into the trap of making it either a 'natural' space or a 'social' space. As discussed above, the preoccupation with emancipating space from the straightjacket of mathematics has a tendency to downplay the significance of the process through which technologies informed by mathematics have been employed in order to establish a spatial reality which resembles Euclidian notions space. Such processes are captured by analysing so-called modern cartography as a particular knowledge technology.

Having introduced a notion of *reality*, the need to *historicize* the *nonhuman*, as well as the notion of *circulating referents*, Latour has paved the way for reading the history of cartography as the history of different practices in which space is established as something real. Mostly focusing on cartography as it has been employed in Europe, the transition to an early scientific cartography beginning in the early fifteenth-century plays a significant role because the shift in cartography altered the reality of space. It is important to note that this alteration, historically, was not a question of cartography alone, but also the technologies that made cartography possible, such as technologies of calculation, transport, measurement and so forth.[9] Nevertheless, a narrow focus on cartography makes it possible to trace the agency of map-making and hereby it shifts the attention away from general technologies, concepts and understanding towards more specific traces of who and what is doing what and when. A historical sociology of space, then, should invoke space as a 'background' framing possibilities, yet, without being a static frame, but one that is transformed as part of the processes we investigate. At the same time, space is also an object, and a means, of power. If the territorial state indicates a state which is defined in terms of a territorial space, then space has to be established as a phenomenon, otherwise it would not be able to identify anything. To repeat Bartelson's argument, space 'turns political' in its recognition as fact (1995: 31). Consequently, I set out to trace cartographic practices which establish space as a phenomenon that allows the state to be identified – and identify itself – by means of this very space.

While maintaining that space should be studied as an autonomous concept, it becomes possible to investigate its historical interrelationship with the state, and how the transition of the spatial identity of the state has worked in tandem with the history of space proper. As the reality of space established through geometric cartography became increasingly complete, its impact on social practice was accordingly increased. This, to the extent that we have come to live in a cartographic reality of space where it is space, as it appears on the map, which forms the basis of socio-political organization. By implication of these arguments, this cartographic reality is not a layer representing or misrepresenting 'an original' natural space, but should rather be considered a specific constitution of reality in its own right. It is worth noting that the cartographic establishment of a spatial reality depends on an ability to make space mobile. This is a profound challenge to the critical geographers, such as de Certeau, who argued that scientific representations work to make space static (Massey 2005: 21). The irony here is that what is supposed to be stable relies on a radical mobility. If this is true, then so-called static state territoriality relies on an ability to render space mobile and, as Latour argues, any seeming stability will not rest on inertia but has to be continuously invoked (1999: 155). Neither a 'state of warfare', nor 'mode of production' or 'class' relations can explain such a spatial settlement and we therefore need to investigate specifically how the reality of space is established historically.

4
The Cartographic Foundation of Territory

Much has been written on the relationship between the map and what it supposedly represents. In *On Exactitude in Science* Jorge Luis Borges famously wrote about an empire where the art of cartography attained such perfection that the map of a single province occupied the entirety of a city, and the map of the empire occupied the entirety of a province. Over time these maps were no longer satisfactory and the cartographers' guilds created a map of the empire whose size was that of the empire and which coincided point for point with it (Borges 1998: 325). This story questions the relationship of representation and the quest for accuracy of the scientific map. Such questions further raise questions about the relationship between the map and the territory. More than 50 years ago, Alfred C. Korzybski stated that a 'map is not the territory' (Korzybski 1948: 58), while half a century later David Turnbull's *Maps are Territories* (1993) sent the opposite message; and emphasizing the constitutive power of maps, Jacques Revel has suggested that 'knowledge of the territory is a production of the territory itself' (1991: 134). Suggesting a temporal diagnosis to this issue of representation, Baudrillard stated that the 'territory no longer precedes the map [...]. Henceforth, it is the map that precedes the territory' (1983: 2). This statement is often used to illustrate characteristics of a supposedly postmodern time in which any substantial reality has been surpassed by representations of representations. In Baudrillard's account we have entered a time where it is no longer possible to distinguish between the real and the imaginary: simulation threatens the distinction between true and false. What we believe is real is in fact simulation: the hyperreal (Baudrillard 1983: 23–5). Hence, the association between postmodernity and a crisis of representation, to use David Harvey's term (Harvey 1989: 260–2), characterized by antifoundationalism where all secure anchorage is dissolved.

These quotes raise two issues concerning the relationship between mapping and territory. The one concerns the role of the map in constructing the object that it supposedly represents; the other is a question of temporal diagnosis regarding the relationship between the map and the territory. If cartographic practice can be understood as a practice that establishes a spatial reality then cartography is not only an issue of representing territories but rather about the spatial conditions for establishing territory in particular, and for the political organization of space, in general. And as geographers have pointed out, different societies produce different conceptions of space and time (Harvey 1996: 443), and almost all societies have produced maps in one form or the other, and in that respect, different modes of map-making are linked to different spatial realities. This means that the map is not only constitutive of territory but of space as a general concept; and in that respect different modes of cartography condition possible ways of organizing territory. This means further that there is no spatial reality outside the map; there is no natural foundation below and besides the cartographic reality of space that we can fall back to.

In sequence, the main argument presented in this chapter, is that the cartographic transition that took off during the European Renaissance provided the spatial conditions for locating sovereignty within a territorial space; that is, defining sovereignty in territorial terms. And this leads to the other issue; concerning historical change. 'Mapping' and 'space' are frequently used metaphors for knowledge production in general. This is especially so with regard to critical accounts of knowledge production where modern knowledge production is sometimes characterized by a belief in a linear relationship between sign and signifier, between representation and what is being represented. Postmodern writings have emphasized how representations are constitutive of objects but, as will be shown, scientific cartography has preceded and played a performative role for territoriality since the fifteenth century. The key to 'scientific cartography' is that it allows 'the map to precede the territory' (King 1996), and while this is not always the case, the principles and prescriptions that constitutes scientific cartography allows for the precedence of the map. And this is no postmodern occurrence; on the contrary, it is essential for the constitution of a modern political spatiality. Thus, like accounts of globalization, which pose an epochal break between a modern territorial geography and a postmodern global politics, so this notion of a present rupture 'where the territory *no longer* precedes the map' is problematic. To use a Latourian phrase, the remainder of this book will strive to open 'the black box' of cartography; the present chapter maintains a more synchronic focus, despite a few historical diversions, than the following two chapters where the diachronic analysis will be unfolded in more detail.

Cartographic histories

We are so used to maps in our everyday life that most people simply take them at face value and few think twice about their impact. The common perception of maps is, indeed, that they are geographically accurate representations of space which help us to find our way and orientate ourselves in space, and they are taken for given to the extent that they are naturalised. Conventionally, the historical map was considered a curiosa, and as an object of academic study, maps were relevant only for historians not least those of cartography. Maps were considered as 'representations of things in space' (Edson 2001: 1899); the classic historical cartographer R. A. Skelton, for example, defined the map as: 'a graphic document in which location, extent, and direction can be more precisely defined than by the written word; and its construction is a mathematical process strictly controlled by measurement and calculation' (1965: 1). The history told about cartography was a linear narrative about increasing scientific progress; in the words of Gerald R. Crone:[1] 'the history of cartography is largely that of the increase in the accuracy with which [...] elements of distance and direction are determined and [...] the comprehensiveness of map content' (quoted by Harley 1987: 3). In response, however, 'a new cartography' has emerged. 'Critical cartographers' have begun to approach maps as culturally specific artefacts, rather than a scientific product and, in this way, maps are given their own prominent place which is attracting interest far beyond the traditional discipline of history. While focusing more on processes of map-making rather than specific maps in themselves, critical cartographers have turned the focus to the 'performative power of maps', that is, how maps are not only representing a geographical reality, but they are serving to shape this very reality.

The power of maps, then, has become a field attracting increasing attention (Wood 1992; Harley 2001b; Jacob 2006; and Edney 1997). Especially, the multivolume 'History of Cartography' has set a new standard for how cartographic history is considered by not setting the scientific map as a natural benchmark against which to compare all other map traditions (Harley and Woodward 1987; and subsequent volumes). In that respect it is sensitive to the poststructuralist and postcolonialist critique of knowledge as being relative to, and intertwined with, power relations rather than being neutral provisions. In the words of Mignolo, '[i]t is needed to get away from the evolutionary model at the basis of the consequences of literacy thesis and to learn from comparative studies and from cultural coevolution' (Mignolo 1995: 323). By releasing map-making from the straightjacket of an unambiguous 'science'. Harley and Woodward's definition promotes an understanding of mapping as being a cultural practice that is neither natural nor universal in its claim to validity. They define maps as 'graphic representations that facilitate a spatial understanding of things, concepts, conditions, processes, or events in the human world' (Harley and Woodward 1987: xvi). This definition has become standard among many scholars who study cartography today, reflect the concern with 'maps as artefacts and with the way maps store, communicate, and promote spatial understanding' (Harley and Woodward 1987: xvi). As such, critical cartography counters the significant Eurocentrism which is prominent among the traditional historians of cartography, such as Skelton who points out that the history of the map is concerned with measuring the 'rate of cartographic progress'; and that it involves the study of 'scientific conquest of the unknown' (quoted by Harley 1987: 3). Reading the *History* ... there is evidence that virtually all cultures have produced maps of their spatial locatedness and, hence, maintaining that the concept of the map has different meanings in different cultures.[2] And abandoning the scientific criteria for what a map is, critical cartography dissolves the hierarchy between the so-called modern map and various forms of so-called pre-modern maps, be they pre- in a temporal sense (for example, medieval maps) or in a spatial sense (maps that stems from cultures other than that of Renaissance Europe).

The emphasis on the power or performativity of maps have inspired a range of theoretical contributions to understand mapping and social space (Cosgrove 1999; 2003); the role of maps for the historical development of states and empires (Black 1997, 2000; Brotton 1997; Buisseret 1992; 2003; Escolar 2003; Biggs 1999; Neocleous 2003; and Padrón 2004) and especially postcolonial-inspired approaches have picked up on cartography as a powerful practice of worldmaking serving European interest in the encounter with other peoples (Craib 2004; Edney 1997; and Mignolo 1995). Thongchai Winichakul has written on the role of cartography for developing national identity (1996; see also Anderson 1991), as has Richard Helgerson in a longer historical context (1992). Studying material culture, Chandra Mukerji has emphasized the role of cartography for the development of a particular understanding of territory (1997; 2006); Joe Painter (2008) has adopted a notion of cartographic reason, or anxiety, in the discussion of regions; and finally David Turnbull has used cartography as a key to understand different knowledge traditions (1993; 1996; 2000). Others have analysed more specifically the role of cartography for the practice of international relations (Henrikson 1975; 2002; Crampton 2006), and John Pickles (2004) has written on the history of spaces through cartography.[3]

All these writings constitute a body of literature too diverse to be easily categorized; yet what they all share is a notion that cartography plays a constituting role for spatial conceptions and practices. And this is, to a degree, to ascribe some measure of agency or causality to cartography with regard to the relationship between social practice and space – even if it is not articulated in these terms. To give agency to maps fits well with the Latourian perspective that ascribes agency to things as well as people, and indeed Harley considered the map an agent of change (Pickles 2004: 48). Drawing on poststructuralist writings generally and Michel Foucault in particular, Brian Harley has been instrumental in conceptualising the power relations inherent in map-making. His writing is concerned with cartographic knowledge as a specific form of knowledge which he reads as 'the discourse of maps' (Harley 2001b: 53). Regarded as discourse, maps pose questions about readership, authorship and the nature of political statements. These are all questions that relate to the notion of discourse as discussed by Michel Foucault, whose engagement with discourse in *The Archaeology of Knowledge* moves the emphasis away from the individual 'author' and representations 'concealed or revealed in discourses' (Foucault 2002: 155). Instead, Foucault asked us to analyze discourse as practices that obey certain rules, and emphasize the types of rules rather than the work of the individual. Translated into the world of cartography, this method implies that we should engage in the study of the 'rules of mapmaking' rather than individual maps and their makers.

Harley is especially concerned with the silences of maps pointing to the things omitted and, hereby, he reads maps as implicitly and explicitly supporting an existing political order. He argues that the discourse of the map has both a scientific and political-social form, which he describes as a strategic silencing and secrecy (power-knowledge), and an epistemological or unintentional silencing (episteme) present in the map (Harley 1988: 279). Because, he is so focused on the things kept secret and silenced by state cartography and reads the power relation as ideology, he is sometimes accused of promoting conspiracy theories against the state (see Laxton 2001). Especially, his notion of intentionality poses questions about the sovereignty of the subject and the degree to which one can ascribe intentionality to the author of a map/text in creating and supporting 'an ideology of the state'. However, if one reads a more generalized notion of power into his theory, one that implies a less predetermined notion of intentionality, we get a fruitful framework for understanding the power of maps as involving two dimensions: one of episteme and one of authorship. The former concerns the general prescriptions that 'scientists' follow in order to establish knowledge. The latter is concerned with authorship, which draws our intention to the question of who is in a position to create maps and who is excluded from this process.

While Harley's dual power relation provides a promising basis for interpreting the relationship between cartographic 'space-formation' and the wider concerns of the spatial identity of the state, an important qualification has to be made. In Harley's definition maps serve a mediator between 'an inner mental world' and on 'outer physical world' (1987: 1). In contrast, in the Latourian understanding, maps should be understood as mediators that translate the goals of actors. It is exactly due to its role in establishing a reality of space, as discussed in the previous chapter, that the scientific map can be said to have preceded the territory and provided the conditions for what is known as 'modern organization of political space'. The Latourian notion of mediation is thus more substantial in terms of the subject/object problematique than that of Harley's which maintains geographical space as a natural constant which is, nevertheless, given meaning in different ways, and used politically, in different societies. Nevertheless, Harley also seeks to establish maps as an agent of change. That is, there is something about maps that has an impact on the way in which social processes occur. In that sense, Harley's perspective is not incommensurable with the Latourian take on cartography as a practice that establishes a reality of space. The difference lies in the subject-object divide. Short-passing this divide, it would be necessary for Harley to inscribe a larger degree of causality to the conditioning power of cartography while the political ideological dimension would have to be downgraded. To give an example of how cartography as a specific form of knowledge is playing a part in power relations, according to Harley, the map surveyor does not only replicate the landscape but also the territorial imperatives of a particular political system (Harley 1988: 279). This is true, of course, but it should also be recognized that the way in which cartography establishes space as real conditions the shape of any territorial imperative to begin with.

Stressing authorship relations invites a historical analysis of who drew the world and for what purpose. This emphasises the focus on agency involved in the establishment of a specific cartographic reality of space. The struggle over authorship has, as we will see, both been a pertinent feature of state formation in Europe, but also in the 'cultural encounter' where different cartographic cultures meet. During the fifteenth century, cartography became increasingly important for the attempts to dominate trade routes vital for the emerging empires. During the sixteenth century, cartography and spatial planning became increasingly significant for the European states in efforts to defend and organize their territories. Now, apart from a couple of general observations concerning authorship, this will be left for the following chapter, and I will focus on the epistemic power relations in the subsequent exposition.

In *The Genealogy of Morals*, Nietzsche highlights the power of the master to name things, and thus sees language as an expression of power (2003: 11). The power of naming can be considered the power to decide the shape of what exists, or in other words, to decide the form that reality takes. This is what Harley calls the discursive power of maps related to the concept of an episteme. In Foucault's definition, the episteme is 'the total set of relations that unite, at a given period, the discursive practices that give rise to epistemological figures, sciences, and possibly formalized systems […]' (2002: 211). In other words, the episteme is a set of relations that provide the conditions of possibility for various sciences. While Harley was concerned with the episteme in order to distinguish the intentional and non-intentional 'silences' of the map (2001c: 87), it also points to paradigmatic differences between map-cultures, or modes of mapping. This power relation becomes obvious in the transition, and in the encounter, between two cartographic epistemes. When we have a transition from one tradition to another, the differences emerge, such as the case with the development from a medieval to a so-called modern map tradition. The other case is the cultural encounter, where two different traditions meet each other, such as is generally the case with the European mapping of the rest of the world. While the latter case has been taken up by postcolonial-inspired studies,[4] I will stress the former (to the extent that they can be separated).

The 'epistemic form', then, decides what 'kind' of space a certain form of map-making produces. Regarding the transition outlined above, I consider this as a shift from one epistemic mode to another which establishes a spatial reality very differently. It was a transition from a cartographic mode based on experience and tradition, to one governed by mathematical principles. As will be discussed in more detail, early scientific cartographic enterprise seems to have been motivated by an ethos of truth, of finding the true way of depicting the world. Only later did cartography become an integrated aspect of statecraft, '[b]y the sixteenth century literary censorship of various kinds was a common aspect of European culture as the emergent nations struggled as much for self-definition as for physical territory' (Harley 2001c: 88). Maps were censored and protected in the same way as written texts. Just as the printing press facilitated a much wider dissemination of maps, some states determinedly kept their maps secret through prohibiting their publication (Harley 2001c: 89–91). Bureaucratic systems of monitoring were set up in countries like Spain, Portugal in the sixteenth century, and by the Dutch East India Company in 1602 (Harley 2001c: 91–6). The secrecy spurred an internal institutional debate over the role of patriotism in scientific argument and the role of secrecy in the growth of knowledge, and Harley concludes that 'access to knowledge must be regarded as one of the more complex socio-legal dimensions that structured the development of cartography in early modern Europe' (Harley 2001: 97). This is the historical context in which the epistemic rules of the discourse of the geometric map evolved.

Cartographic transition

The cartographic representation of the earth changed dramatically in Europe during the fifteenth and sixteenth centuries, and the transition from a medieval to a scientific mode of mapping marked a turn from a tradition of mapping 'space and time' to one oriented towards the spatial present. Conventionally, medieval maps have been regarded as being imbued with a sense of religious superstition and inadequate spatial understanding. Yet, as argued by Evelyn Edson, medieval maps served a different purpose than the scientific modern map. Rather than representing the presence of space according to abstract geometrical principles, these maps generally aimed to organise space according to religious and philosophical principles. Different traditions and principles can be identified in medieval cartography, but one of them, the tradition of *mappaemundi*, aimed to portray universal history as well as historical space (Edson 1997: viii, 15). These maps generally had an encyclopaedic character as expressing various 'kinds' of knowledge. Their representation of time and space was relative and they appeared as the only medium that could represent the multidimensional reality of European medieval life (Scafi 1999: 63–4). They enabled the visualization of a spatial paradox in that the same map would show the divine Garden of Eden as well as earthly locations such as Paris, Rome and Jerusalem (Scafi 1999: 53).

The *mappaemundi* generally do not show signs of authority other than that of Jesus. There are no territorial divisions and there are no signs of territorial demarcations. These maps were not useful for navigation, travelling, planning military campaigns or negotiating treaties. But they were useful for contemplating the nature of the universal order and the symbolic significance of places in a temporal world. They supported 'the central medieval utopia of united, universal Christianity, the republic of Christ, under the leadership of the Catholic Pope and the Holy Roman Emperor' (Jespersen 2004: 13). In that respect, this tradition was locked in a specific symbolic ordering of the world. Just as other cartographic traditions, such as Chinese or Aztec cartography, the *mappaemundi* were centred on the most significant locations within its own world. In effect, the principles informing this map-tradition could not be 'dis-located' and used to map another part of the world, as the social content of the map would lose its meaning.

From approximately the time of the Hereford map (see Map 4.1), another way of mapping emerged in Europe. The *portolans* were primarily a Mediterranean mapping enterprise carried out by merchants and sailors. They are known from the late thirteenth century (Parry 2000: 101) and they were used for navigation, describing compass bearings, dead reckoning, landmarks and coastal features. In that way the *portolans* were a typical 'experience' based rather than a theoretically or conceptually structured accumulation of knowledge. As a reflection of their purpose, they depicted coastlines, cities and place names along these as well as rivers. Visualization in this tradition was combined with textuality as the *portolans* were accompanied with detailed textual descriptions of travel routes. Initially, they were used for navigation but eventually the information and style of the *portolans* were integrated with the *mappaemundi* tradition (Edson 1997). A prime example of this is the Catalan Atlas which was made in 1375 for King Charles V of France. It contains an entire cosmography and geography on six different tables all together describing the state of the known world (BNF 1998). This merger between practical symbolic ordering and practical purposes could be seen as a pre-warning of the transitions which would occur during the fifteenth century.

Between the fifteenth and the sixteenth centuries, Eden disappeared from the world map, which points to a 'shift from medieval to modern thinking from a holistic to a fragmented view of reality, from a mapping which sought to penetrate mystery of the whole universe beyond human boundaries to a mapping which is contained strictly within the frameworks of analytical thought and Euclidian geometry' (Scafi 1999: 70). The location of Eden was increasingly at odds with experience, and it was incommensurable with the epistemic foundations of the new

Map 4.1. The Hereford Cathedral mappaemundi, ca. 1300. The map is orientated with East at the top and Jerusalem at the centre. Europe and Africa are the bottom two continents on either side of the Mediterranean. The small circle on 'the top of the world' in the top periphery of the map represents paradise. A number of the most significant towns are depicted as well. Courtesy of Hereford Cathedral.

cartography that came to be based on abstract mathematical principles as part of a general move in the conception of space. The combination of Ptolemy's 'Geography' and Euclid's 'Geometry' epitomises the modern conception of space as being abstract, homogenous and, importantly, separated from time.

It is common among those that discuss the new 'culture of space' in the Renaissance to draw on Samuel Edgerton's study on the rediscovery of linear perspective (for example Ruggie 1993; Walker 1993), where he argues that the introduction of the linear perspective is related to the influence of Ptolemy's *Geography* in Florence from the first decade of the fifteenth century. 'Ptolemy's conception of the earth as having its surface organized by a grid system of longitudes and latitudes, so that all parts could be thought of in proportion to one another' (Edgerton 1975: 111) provided inspiration to the technique of representing objects in space according to the linear perspective. As argued by Anthony Blunt, this encompassed a new approach to the world; an approach which was first and foremost about rendering the outside world according to the principles of human reason (1968: 2). In the words of Leon Battista Alberti: '[t]he function of the painter is to render with lines and colours, on a given panel or wall, the visible surface of any body, so that at a certain distance and from a certain position it appears in relief and just like the body itself' (quoted by Blunt 1968: 14). What is striking about this quote is the emphasis on lines, colours and bodies in the exact imitation of nature, and generally the insistence on geometry and mathematical unity for the 'new perception of space' (Edgerton 1975: 113).

In cartographic history as well, the translation of Ptolemy's *Geography* into Latin during the first decade of the fifteenth century is frequently named as the symbolic beginning of this process because it (re-)introduced the principles that inform scientific cartography to Western Europe. However, the extent to which this was the reason of Ptolemy's popularity is questionable. In a recent study Patrick G. Dalché rejects Edgerton's thesis, and suggests instead that Ptolemy's *Geography* served to codify and inspire an existing debate concerning space and the organization of cosmos. Initially it was not the technical prescriptions for cartographic representation that gave the *Geography* its status as an indispensable reference but rather the desire to understand the world of the ancient authors admired by the Renaissance scholars (Dalché 2007: 287–98). Though, even if it was a desire to understand the world of the ancient sources of learning that provided a motive force behind Ptolemy's popularity at first, the discovery of lands unknown to those writers created an incentive to modify the world map towards the end of the century. The encounter with a hitherto unknown continent changed the known scope of the world and shattered the medieval cosmography dividing the world into three continents: Asia, Europe and Africa (Jahn 2000: 22–50). This might well have contributed to creating a more receptive environment for the principles and technologies codified by Ptolemaic cartography, which contained coherent prescriptions for representing all localities in relation to a graticule (that is, a grid of parallels and meridians) (Ptolemy et al. 2000: 31). Ptolemy had worked out different projections which would, in effect, give a different shape to the space represented. Hereby, he provided the principles for a representation of space based on theory rather than on experience.

It was an episteme that contributed to a 'science of measurement' corresponding to what Foucault has described as the classical episteme of science characterized by a 'universal science of measurement and order'. and the 'principle of classification or ordered tabulation' (Harley 2001c: 97). In addition to Ptolemy's *Geography*, this drive towards scientific map-making was significantly informed by Euclid's geometry bearing on a concept of abstract, geometric, homogeneous space (Brotton 1999: 74). In the *Elements*, Euclid presented a small number of axioms from which he deduced hundreds of propositions, for example, his famous definition that '[p]arallel straight lines are straight lines which, being in the same plane and being produced indefinitely in both directions, do not meet one another in either direction' (Morrow and Euclid 1970: 137). Euclidian geometry treats space as a void in that the rules that govern space are based on abstract mathematical principles formulated in a universal language that refrains from taking into account bodies or matter that may exist in space. Space in this system is conceptually ordered by a grid system of latitude and longitude, and this makes it possible to imagine the globe as an empty, or blank, space only covered and subdivided by intersecting lines of latitude and longitude. Secondary to this ordering, spatial observations can gradually be added, or written into this grid system in order to complete the cartographic image of the world.

One thing, nevertheless, is the transformation of the ideas and principles of cartography, but on a more general level, advancement in calculation techniques and practices also made this possible. To draw maps according to the new principles it was necessary to be able to observe and measure specific locations in relation to celestial objects, to measure time differences as well as calculate and combine these observation with existing 'data'. It was necessary to be able to apply geometry on the ground, so to speak. The measurement of latitude had always been the easier one to do. By the middle of the sixteenth century there were two well-established methods of calculating this. Both were concerned with measurements of the angle of the height of celestial bodies above the horizon: the sun and the polestar. They were measured with variants of the astrolabe which had been known also by the Greeks and the Romans (Brown 1949: 180–1). By measuring this angle it was possible to determine the distance from the equator. Longitude, on the other hand, was much more complicated to measure due to a lack of precise technological equipment, especially, the absence of an accurate clock. At sea, the distance sailed was measured by 'dead reckoning' (calculating distance by multiplying the estimated speed of the ship with sailing time) which, combined with other knowledge such as the measurement of latitude and knowing the direction sailed, made it possible to determine an approximate location of the ship. Over the years, various suggestions of how to determine longitude at sea were conjured up, but, until a reliable way of measuring time was invented, there were no practical methods of calculating longitude precisely (Brown 1949: 209–11). On land, it was with the surveying technique of triangulation, whose practical prescription is accredited to Gemma Frisius in the first half of the sixteenth century, which made it possible to calculate distances mathematically by measuring angles between features in the landscape and relating them to a measured baseline. One of the first known systematic uses of this technique for local area mapping was carried out by the astronomer Tycho Brahe who measured his island Ven in 1578 (Ehrensvärd 2006: 103). Later, triangulation provided a basis of the great mapping project by the *Académie Royale des Sciences* in France in the second half of the seventeenth century. The question, now, is what this transition of the ideas, theory and technology of map-making meant for the political organization of space and the development of the territorial state.

A cartographic reality of space

It is important to note that the cartographic transition was not a matter of a translation of a single volume but was part of wider developments in art, science and politics during the European renaissance. The character and causes of these changes have been subject to discussion and it is clear that the transition towards geometric cartography did not take place independently of these. The period is generally considered to represent great changes both in terms of material conditions, and the apprehension of the world. In one account, three principal trends generally characterize this period in Europe: first, the recovery of populations and the economy after the Black Death had devastated communities across Europe contributed to a growing wealth of societies; second, the rupture of Christendom weakened the Catholic church, questioned its universality and opened space for other types of authority dominating, such as secular state authority; third, the foundation of overseas empires started the process of integrating the entire globe into a European economic and political system (Brady et al. 1994: xvii). This last trend, in particular, entailed a novel way of regarding the world. It was within this context that cartography developed, and by strictly emphasizing projection, location and calculation, the focus was turned to the present while historical narratives were written out of the map. Time came to play an important role in calculating and measuring distances which contributed to a general 'rationalisation of space' (Woodward 1991). Though, reflecting the arguments regarding the autonomy of space presented in the previous chapter; if regarding cartographic change as simply part of a wider transformation, then the specific conditioning power of cartography is subjected to be just that; part of a wider transformation. In order to grasp the significance of cartography for establishing a specific spatial reality it is necessary that a measure of autonomy is granted to cartography as a social practice.

Before geometry became the general standard for spatial measurement, practical usage and units of time were often the standard. In consequence, the representation of space in the Middle Ages did not provide a knowledge that allowed for sharp boundaries to be drawn or that enabled the uniform organization of large territories. Medieval political territoriality is usually described as being overlapping and hierarchical with no clearly defined centre and with ambiguous boundaries. Territory did not play the same role in defining the domain, but followed rather as a result of 'jurisdictional sovereignty' determining the relationship between subjects and the ruler (Sahlins 1989). Control of the territory was generally maintained by controlling the towns and, not least, the castles of the country. This was achieved by direct ownership or via personal relations of allegiance between the king and the lords. Such relations were generally personal and had to be renewed with the heirs when the noble man died. Control and knowledge over the territory was thus achieved through a web of personal intermediaries.

With the advent of scientific cartography, space was established as an 'autonomous' category, which, in principle, could be perceived without reference to functional time, the immediate experience of the environment, or the author of the map. Instead, locations became locations *in* space and space became a matter of the relationship between celestial features and the Earth where distances were calculated by use of degrees, triangles and geometry. Generally speaking, the 'mode' of knowing space changed from one based on literacy and tradition to one of mathematics and visual representation (Turnbull 2000). This new way of representing space enabled a uniform visualization of territory which, so to speak, rendered it coherent and tangible in its own right. Thus, being emptied of any other substance than its own, it became possible to 'write things and features into space' which did not necessarily exist. For example, it became possible for princes and emperors to assert their authority over a territory without the need for intermediaries even if they did not control it. In that respect, the scientific map also rendered space as a 'thing' which could be possessed – not as a field, a forest, a mansion, but as space. Based on the spatial reality established by the scientific map, it became possible for two neighbouring rulers to sit over a table and agree (or disagree) on boundaries and draw them on the map. By means of the grid system, this boundary could then be located on the ground independently of what existed there prior to the agreement. The cartographic transition also broadened the scope of space 'possibly within reach' of the state apparatus. Striving towards establishing a centre of calculation under its control, the state would increasingly be able to produce a territorial space. Generally the scientific map expanded the tax-base of the territorial state in the countryside and improved its ability to collect taxes, and the defences of the territory were pushed to the boundaries as the cartographic image of a coherent territory was slowly made real on the ground.

In a paradoxical way the geometric map makes place appear less important because it designs all places to look the same and, in that way, strip space of its social context (Harley 2001a: 167). This abstraction of space involves a separation of time from space, and the production of a specific form of space. The 'universalisation of space' – or isotropic conception – means that neither spatial nor temporal differentiation affects the measurement and representation of space. It is important to note that 'time' as such did not lose significance for the calculation of space. Most obvious is the case of longitudinal distance which is measured in terms of time difference. However, time was separated from space in the sense that it disappeared from the representation and immediate metronomic and, thus, the fabrication of space *per se*. The temporal and spatial order created by time zones is in effect a separation of time from space and a subsequent reinscription of time into an abstract spatial framework of modern cartography. This distinction between space and time in the map is only a valid claim according to the principles of the map itself. These principles, as well as the actual practice of mapmaking, are never free of more general spatial and temporal constraints. Yet, even though a map is never independent of when it is produced, the cartographic image itself appears with no reference to time (maybe apart from an inscribed date of print and a dedication to a patron – see, for example, map 6.4, p. 137).

The abstraction of space in its cartographic form detaches the representational image of space from any bodies that may occupy this. Thus, the epistemic principles of the scientific map produce space as empty. This is to say that the scientific map dissolves the symbolic centre of authority which was the governing principle of the medieval *mappaemundi*. As we will see in the subsequent chapter, to produce maps of one's world centring it on a symbolic centre of authority has been a common feature throughout the history of cartography globally. Dissolving any pre-defined centre represents a particular dimension of power of the scientific map because it renders this particular mode of mapping space *movable*. To exemplify this, it would have been impossible to take the principles governing the *mappaemundi* and then apply them to a mapping of Australia unless this was done in relationship to Jerusalem as the navel of the world. The principles defined what space could be mapped. The scientific episteme allowed map-makers to map all areas equally in the sense that there was no pre-defined centre of the map. According to the scientific principles it would not matter whether you were mapping the Arctic, France or South America, and this opens for the possibility that other cultures could adopt scientific cartography as a way of mapping while maintaining their own privileged position in the centre of the map. This is not, however, to say that this is a neutral way of establishing space, and in practice European map-makers maintained Europe in the centre of the map and prioritised Europe's location in the world. The fact that geometric cartography is compatible with different value systems does not mean that it is value free, and as has been pointed out by postcolonial scholars, the linguistic and conceptual acquisition of space has been a core feature of colonialism (Mignolo 1995; Cohn 1996). Nevertheless, it is very important to understand the impact of the scientific map dissolving the symbolic centre of the map, and hereby making this mode of cartography mobile beyond its initial location of development.

Ignoring thus for a moment, that all universals necessarily will have their origin in a particular context, it becomes possible to speak a universal language of space in which the same spatial coordinates (degrees, minutes and seconds) signify the same location for everybody. The abstraction from the social actuality of places thus facilitates large-scale planning and reorganization of vast territories. Because the space established by the geometric map is disassociated from the people that occupy it, it becomes possible to treat it as if it were empty. This abstraction of space is also considered as a rationalization of space in that the principles of representation create a universalized rationalized order of space and time that makes rational planning and coordination possible. In the words of David Woodward, space was created as abstract, homogenous, and geometric, which fed into a cultural turn 'around the idea of measured space' (1991: 84). This rationalized image of cartographic space can always be written into different contexts, and inform various imaginations born out of different contexts because, again in the words of Woodward, the world was represented as 'a *tabula rasa* on which the achievements of exploration could be cumulatively inscribed' (1991: 85). Writing space as blank enabled states and corporations to create spatial entities that were subject to their power; in the words of Jouni Häkli, 'it is precisely the capability to visualize a society as both an object of administration and a subject of politics that has made the efficient and rational functioning of the modern state apparatus possible in the first place' (1998: 134). By invoking a triple abstraction separating space and time, space from bodies, and space from the observing subject, the map establishes space as coherent, yet, empty, and this provides the possibility of writing and rewriting meaning into space to a degree that exceeds that of medieval and related map epistemes.

To summarise, the transformation of spatial knowledge transformed the spatial reality in a way that established space as 'an autonomous phenomenon'. This, in turn, made it possible for space to determine other social relations in novel ways. It not only enabled novel forms of control, demarcation and abstract planning in, and of, space, but it allowed space to become a social determinant in a way which would not otherwise have been possible. The consequences of geometry coming to inform the representation of space combined with a new notion of science implied that mapping should be a scientific representation of an independent existing physical reality. This presupposes a distinction between the observing subject and space and the isotropic conception of space implies that, in principle, all space can be measured according to the same principles. That is to say that the scientific principles of map-making present themselves as a universal application for representing space and objects in space. The employment of such a set of principles implies that space can be represented independently of the bodies that occupy this space. As a result of the epistemic rules of maps, space is abstracted from its author and observer, it is abstracted from spatial (geography) and temporal (history) differentiation, and abstracted from bodies and matter, or what Lefebvre (1991: 131–2) calls texture, and what Sack (1980: 4–5) calls substance, of space. Crucially this transition allowed sovereignty to be defined and demarcated in spatial terms, thus conditioning the possibility of the sovereign territorial state.

Cartographic territories

In order to establish a direct link between sovereignty and territory it was necessary to render territorial space somewhat autonomous from social practice. In order for the relation between territory and sovereignty to become a determinant of other social relations it was necessary that it obtained a seeming existence relatively independent of the social relations that sovereign territory was supposed to define. As discussed, the map mediates a certain reality of space, and space is established as an autonomous category *as space*. Space in this scientific schema was no longer dependent on occupying bodies or symbolic meanings but was established on its own terms. As Latour argues, by being established as real it is established as autonomous. Emphasizing the significance of geometry, Stuart Elden quotes Paul Virilio: 'Geometry is the necessary foundation for a calculated expansion of state power in space and time' (Elden 2005a: 14), and concludes that '[t]erritory in the modern sense requires a level of cartographic ability that was simply lacking in earlier periods, an ability that is closely related to advances in geometry' (Elden 2005a: 15). The key for this argument is the way in which this transformation of space informed by geometry specifically entered a new relationship with the state power.

As a representational practice, the map not only constitutes its own object it also necessarily enters into a system of relations with other representational practices and, in so doing, alters the meaning and authority of others (Helgerson 1992: 146–7). Maps usually show more than just the physical contours of the landscape; there is typically a reference to the patron and an institution whose authority is behind the map. In terms of political maps, administrative and state boundaries are usually marked on the map, and this is a clear carto-political act. Boundaries are agreed (or enforced) by mutual recognition between collectives, and these boundaries cannot be based on the strict calculative measures informing the episteme of the map but has to be based on a political decision or agreement. With the transition to a geometric cartography and the gradual assemblage of space as an autonomous and abstract category it became possible for political decision making and administrative practice to refer to the cartographic representation as a real representation of the territory. In 1440, for example, a longitudinal boundary was suggested in the conflict between Florence and Milan and this was possibly the first instance of imposing an imaginary mathematical line as a political boundary in Europe (Edgerton 1975: 114). Hence, the way in which space is made real through cartographic practices, in a very general sense, affects the way in which political decision and practices can refer to this space, and therefore, when inevitably blending into the politics of the state, cartography alters the way in which boundaries can be invoked and conceptualized. And in doing so, the map provides a spatial identity to what is represented on the map. In effect, what I suggest is that the scientific map alters the conditions of representing political identity and that the map constructs its specific image of territorial space as being political space.

As the representation of space was released and abstracted from moral conventions and social functionality, it was, then, acquiring primacy as an ordering principle. It enables a deliberate politics of space because it constructs territory as empty, which allows potentially any substance, any identity to be written onto this space. In that respect, the construction of territory as cartographic space is the condition of possibility of the writ of boundaries by a sovereign authority onto a society. With geometric cartography, it became possible to establish boundaries as 'sharp lines' rather than the gradual frontier zones and ambiguous delineations that characterized medieval polities. Drawing them on the map, it was possible subsequently to trace them in the landscape, in principle, by scientific measurement. Without the advent of geometric cartography, it would have been nearly impossible to establish boundaries as neat lines as a characteristic of the state. Without the abstraction and relative autonomy of space, it would not have been possible to give space primacy as a defining dimension of the state, providing it with a clear territorial identity that allowed for self-reference in terms of an abstract geo-body independently of the actual rulers of the state. In the following two chapters this relationship between the map and the territory will be investigated in greater detail when the focus is turned to specific mapping practices and, eventually, state formation.

This chapter has shown how cartography can be seen as a social practice that assembles a specific reality of space. When analysed along two dimensions of power; one of epistemic rules and one of authorship, we get a framework where we can analyse both the conditioning effects of cartographic change as well as the agency involved in map-making. Expanding the Latourian perspective on space formation, this entails that we can trace the agency of space formation and at the same time treat cartography as a *mediator* that translates the environment into a specific assemblage of space. And by analysing the epistemic rules that informs a certain mode of cartography, we can analyse the conditioning power of a specific configuration of space.

In consequence of the epistemic rules gaining dominance with the cartographic transition in Europe during the Renaissance, so-called modern territorial space should first and foremost be treated as a cartographic space, and this cartographic space provides the modern state with its much disputed territorial identity. As science establishes space as abstract, geometric and homogeneous, it dissolves the symbolic content of previous mapping practice, and thus erases any 'pre-defined' symbolic centre of authority. Hereby, the scientific map can, in principle, claim universality and embrace the entire globe, while at the same time establishing this as something that can be readily appropriated and manipulated by people. In establishing space in such a way, the geometric maps provides new conditions for states to create and use space, and hence, lead to the notion that modern state territory is first and foremost a cartographic construction. And hence, the precedence of the map over the territory is not a postmodern phenomenon, but rather a core function of the geometric map and the epistemic rules that inform this. This, in effect, forces us to rethink Baudrillard's statement that 'the territory no longer precedes the map' in present times. He states that abstraction is no longer about abstracting from a real – today it is about the 'the generation by models of a real without origin or reality: a hyperreal' (Baudrillard 1983: 2). Taken literally, Baudrillard's statement logically presupposes a time where the territory preceded the map, hence the map functioned as a representation or abstraction from a 'real territory'.

While maintaining that a significant transition took place, the notion of territory as cartographic space also has implications for the discussions concerning modern vs. postmodern political space and spatial identities. It is not uncommon to see arguments challenging a so-called early modern geographical imagination; for example, that traditional identities are dependent on traditional (modern) knowledge (George 1994: 216), or that identities which do not fit into the contours 'of the modern geopolitical map' ought to be recognized as politically real, thus, existing in global politics (Shapiro 1994: 479–82). Feeding into these arguments, David Harvey's well-accredited argument about postmodernity suggests that the postmodern condition represents a crisis in the representation of the experience of time and space (Harvey 1989: 260–2). However, as I have argued here and in the preceding chapter, maps should not be considered as simply representing space – in its realist or performative understanding, but rather as *mediators* that help to establish a specific reality; that is, the relationship between maps and space is one of construction. This is important because if we refuse the premise of a subject/object distinction as the foundation for our understanding of space than there is no foundation outside the specific assemblage of space. There is no nature of space outside the practices – including both *human* and *non-human*, remember – that constitute space as space. And this is why, essentially, that the spatial conditions for territory are provided through cartographic practices. And, in sequence, if we wish to know whether changes are taking place in the organization of political space, we would have to include an analysis of whether the current practices of map-making are changing rather than suggesting that the map is in crisis based on ideas of relative conceptions of space, or a representational crisis identified in the arts and other areas. Is a new set of epistemic rules coming to define cartography in the supposedly postmodern era?

This, in other words, serves to break away from arguments concerning political identities relying on generalized discourses of pre-, post- and simply modern knowledge. Maintaining the relative autonomy of cartography and space allows me to emphasize the conditioning power of cartographic practices on other practices; and by stressing authorship as the other dimension of cartographic power, I will analyse how cartography blends into politics in a novel fashion following the cartographic transition commencing in the early fifteenth century. The amalgamation of cartography and early modern state power is crucial for understanding the spatial identity of the state. In Helgerson's interpretation, maps are as much a representation of power as one of space because maps indicate sovereignty and ownership of space (1992: 107). This is true, as one can observe in Saxton's maps of Britain from the sixteenth century that prominently showed the insignia of the Queen Elizabeth. It is, however, important to remember that the cartographic transition was not necessarily led by instrumental state power, but rather, provided possible means to a new assertion of sovereign relationship between the land and royal power. In the following two chapters, I will provide a historical exposition of global and territorial mapping, and state formation.

5
The Cartographic Formation of a Global World

Some years after the publication of Abraham Ortelius' *Theatrum Orbis Terrarum* in 1570, a grateful owner of the atlas expressed his admiration in the following words: 'You compress the immense structure of land and sea into a narrow space, and have made the earth portable, which a great many people assert to be immovable' (quoted by Brotton 1997: 175). The praise of this statement contains significant ideas concerning the mobility and unity of global space that speaks to the theoretical concerns raised so far. The purpose of this chapter is to show how the map preceded and produced the globe as a social space or, in other words, how the globe has been unified through cartographic means prior to, and concurrent with, European imperial projects. It will be argued that mapping was prioritized as it enabled the coordination of social practice on a global scale, as if the world was a single and unified space.

In most claims to a postmodern globalization looking to the future, and most often in historical sociological accounts of how we came to be living in the world of nation states, the globe is taken to be a natural outer sphere demarcating humanity, homo sapiens, in general. That is, global space is generally taken for granted as a natural space that both frames social relations and also provides a surface – or a stage – on which social actors perform. Space is thus something we can travel across (Massey 2005: 4–8), or something that provides an obstacle to movement which can be overcome through technological feats, but most importantly, for my argument, is the understanding that the globe is a natural object which provides the frame, and therefore the limit, for social practice. Conceptualized in this fashion, the globe provides a natural unity to humanity – or a social totality – which does not have to be questioned or investigated, and therefore it provides a natural reference for politics. In IR, questions have been raised concerning the globe as a natural container of politics; in *The Expansion of International Society*, for example, Bull and Watson seek to explain the historical origins of the current global international system which, from their perspective, has superseded a situation where the world was characterized by a number of separate regional international systems (1984: 1–9). The question they raise is an important one; yet like most writers they focus on the expansion and encounter of social relations and networks, but assume that the stage on which this is done remains a constant one.

Eric Wolf's brilliant *Europe and the People Without History* illustrates this nicely through his central assertion that the world of humankind is a manifold totality and those analyses that split this totality into bits, without proper reassembling, 'falsify reality' (1990: 3).[1] He emphasizes that humankind has always lived in one world and points to global production systems, such as the Dutch in the Bengal (Wolf 1990: 3–4). These are, he argues, common but largely ignored facts. Drawing on Marx, Wolf asks whether we can envisage a common dynamic of the connections and yet maintain 'a sense of its distinctive unfolding in time and space as it involves and engulfs now this population, now that other' (1990: 19)? In search for a positive answer to this question he embarks on a world tour with an 'imaginary voyager' in the year 1400. Wolf argues that to understand this world, 'we must begin with geography. A map of the Old World reveals certain physical constants' (Wolf 1990: 25) such as mountains running east-west across the Eurasian landmass. Now, the trouble arises in this search for 'physical constants' and the main assertion of the totality of humankind and the singularity of the world as a fact, rather than as something that itself has to be explained historically. Wolf maintains a distinction between nature and culture that histoticizes social relations, but fails to take into account the processes and the agency that produce space as one world. This is to say that space is taken outside 'the social' and appears as a constant background whose shape and size has some causal effect on the formation and modality of social relations. He notes that the voyages from Marco Polo 1271 to Columbus 1492

> were not isolated adventures but manifestations of forces that were drawing the continents into more encompassing relationships and would soon make the world a unified stage for human action. In order to understand what the world would become, we must first know what it was. I shall therefore follow an imaginary voyager in the year 1400 and depict the world that he might have seen.
>
> (Wolf 1990: 24)

Wolf thus recognizes that social relations are not necessarily integrated into a unitary totality and that this occurs as a result of historical processes, but he maintains nevertheless the social totality as an ontological fact. This is nicely illustrated when, in depicting the world of his imaginary voyager, Wolf presents a modern map of the world on which the main trade routes of 1400 are drawn (1990: 28). But when Wolf sends his ship out in 1400 the world map did not look like this and there was not one world map that was universally accepted. If we were to look at world maps in different parts of the 'global world' circa 100 years prior to Columbus' voyage, we would see significantly different worlds – and, more importantly, we would see that the world map was not necessarily a global map.

The notion of natural space as a stage on which life occurs neglects the social processes which have been involved in producing this stage itself. If space is a social relation, then a notion of social totality can no longer rest on a notion of natural spatial unity as an ontological fact. Global space, then, was produced as a 'stage' on which states could 'act'. It is no accident that Ortelius' atlas is named 'theatre'; as described by Erik Ringmar, the metaphors of stage and actor emerged during the Renaissance as a prevailing way of understanding the state and its habitat (1996: 443–6). The metaphors describe how certain polities perform on the global stage, and they also reflect the image of a level playing field on which political actors relate to each other in various ways.

Returning to the happy owner of Ortelius' atlas, a key dimension in the cartographic assemblage of global space was the ability to make space mobile; in order to bring together the disparate sites and places of the world and establish them as a single space it was necessary to establish a method for making these sites mobile. In making this argument, the chapter starts off with the increasing global rivalry between Portugal and Castile in the fifteenth and sixteenth centuries which rendered the globe a political space during the European Renaissance, and then it turns to the epistemic transformation of spatial knowledge encapsulated in the growing discipline of Cosmography that enabled the reality of the world to be established as a global space. This is where techniques and calculations are developed to render space mobile. With the Spanish attempts to create a master map, or general pattern map, the *Padron Real*, such techniques were institutionalized leading towards a gradual completion of a new global world map which I see as being completed with publishers such as Abraham Ortelius and his atlas *Theatrum Orbis Terrarum* typifying the novel world that became a global standard. The final irony here, concerning the notion of mobility and globalization is that from Latour we remember that since social action cannot take place in abstract space; space has to be made present and concrete. Hence, it was the cartographic unification of the globe that allowed 'action-at-distance', that is large scale planning and coordination with a global reach. As such, the apparent static quality of the modern world map rests on a profound mobility. In consequence, the mobility associated with globality in fact rests on an ability to assemble a seemingly stable global space which, in turn, rests on an ability to make space mobile in the first place.

Turning the globe political

In 1453 the Ottomans conquered Constantinople and put an end to the Byzantine Empire. Around the same time, a peace treaty was signed to mark the end of the Hundred Years' War between France and England signalling the early formation of two 'territorial states' outside the Holy Roman Empire, but within the bounds of a universal Christianity. Both events were marked by processes of spatial delimitation which challenged medieval notions of universality. During the fifteenth century the notion of Europe came to be understood as a geographically located Christian republic, and universality was thus increasingly understood as a spatially bound phenomenon (Mignolo 1995: 326). Further, with the defeat of Constantinople, and what remained of the Roman Empire, it was no longer clear where 'the universal empire' was. Spain, under the Habsburg monarchs Charles V and his son Philip II, did pursue ambitions of a global world empire until Philip's death at the end of the sixteenth century. The Ottomans also pursued imperial dreams. In 1466, Mehmed the Conqueror received a letter from a Greek scholar telling him that 'No one doubts that you are the emperor of the Romans. Whoever holds by right the center of the Empire is emperor and the center of the Empire is Constantinople' (quoted by Brotton 1997: 92). Not only did the Ottoman conquest introduce a challenge to the relationship between empire and a universal Christianity, it also affected trade patterns in and around the Mediterranean. In effect, the Ottomans came to control, or mediate, much of the lucrative trade with Asia and this provided an incentive for the West European rulers to seek alternative trade routes.

During the fifteenth century, both Spanish and Portuguese rulers were in competition with each other seeking to establish and control trade routes along the East coast of Africa and the Atlantic islands. In the 1450s, Portugal claimed a monopoly on the Guinea trade on the grounds of being the first to have discovered these places and having received Papal bulls granting this right (Parry 2000: 134). A subsequent war of succession between Spain and Portugal was resolved with the Treaty of Alcaçovas in 1479 containing a general demarcation line dividing the contested areas in Atlantic and East Africa into a Spanish and a Portuguese sphere. Portugal maintained its dominance at sea and Spain refrained from interfering in the Guinea trade. The division produced by this Treaty had implications that led to the Treaty of Tordesillas between Portugal and Spain from 1494 which settled an abstract boundary dividing the world into a Spanish and a Portuguese sphere where they could respectively claim possession. And as such the Treaty of Alcaçovas became a forerunner and an expression of a novel way of conceptualizing and dividing the globe resulting in a *politicisation of the globe*.

The event that triggered the Treaty of Tordesillas was Columbus' expedition to the Americas. Returning from the first voyage, Columbus' fleet was forced to seek shelter in a Portuguese port in the Azores as a consequence of adverse weather conditions. Being suspicious of Columbus' voyage, the Portuguese authorities laid claim to his discoveries under the Treaty of Alcaçovas (Parry 2000: 151). In response, Spain sought Papal support and received it. Four bulls were issued by the Pope, and the third – *Inter Caetera* – drew a boundary line that laid the foundation of the Treaty of Tordesillas that was ratified the subsequent year (Parry 2000: 150–2). In this Treaty it was agreed

> that a boundary or straight line be determined and drawn north and south, from pole to pole, on the said ocean sea, from the Arctic to the Antarctic pole. This boundary or line shall be drawn straight, as aforesaid, at a distance of three hundred and seventy leagues west of the Cape Verde Islands, being calculated by degrees [...]. And all lands, both islands and mainlands, found and discovered already, or to be found and discovered hereafter, by the said King of Portugal and by his vessels on this side of the said line and bound determined as above, toward the east, in either north or south latitude, on the eastern side of the said bound provided the said bound is not crossed, shall belong to, and remain in the possession of, and pertain forever to, the said King of Portugal and his successors. And all other lands, both islands and mainlands, found or to be found hereafter, discovered or to be discovered hereafter, which have been discovered or shall be discovered by the said King and Queen of Castile, Aragon, [...], and by their vessels, on the western side of the said bound, determined as above, after having passed the said bound toward the

west, in either its north or south latitude, shall belong to, and remain in the possession of, and pertain forever to, the said King and Queen of Castile, Leon, [...], and to their successors.[2]

This text is significant for a number of reasons. First, it renders the whole world a space for Spanish and Portuguese expansive trade and conquest – even those places that are as yet unknown. Second, it divides this world according to an abstract line of longitude that does not take into account what already exists in those places that might fall under the auspices of the Spanish or Portuguese crowns. Thus, being based on an abstract knowledge of space, depicting the world as an empty globe where all places can be located according to the grid of latitude and longitude, the Treaty also pre-empts future controversies over spatial possession as authority over new land will simply be a matter of location within the matrix. In this respect it was already decided that Brazil would belong to the crown of Portugal even prior to its 'discovery' in 1500 and it was thus the Treaty based on a cartographic reality of the world that came to decide 'the reality on the ground' and not the other way around. Therefore, in the years to come, legitimate claims to possession became inseparable from determining the location of longitudes and latitudes and deciding the location of coastlines and islands.

The Tordesillas Treaty stands as maybe the first genuine example of how a political boundary originates in a map or rather, in a carto-scientific representation of space which, in consequence, plays a performative role in shaping the world. Politics no longer primarily corresponds with socio-political relations, or 'the reality on the ground', but corresponds instead with a reality derived through cartographic means. To exemplify, in 1515 a group of Portuguese were imprisoned for having violated Spanish rights by landing on the Cape St. Augustin of Brazil, but controversy arose as it was not actually known where this cape was located in relation to the demarcation line. Although the boundary was known, it constantly had to be assessed when confronted with events that turned locations into contested sites. Though, as long as things remained within the framework of Tordesillas, the solution had been rendered a technical one rather than a political one. In response to this particular problem, a 'junta' (of scientists and experts) was assembled to determine whether the Cape St. Augustin was on the Spanish or the Portuguese side of the demarcation line (Lamb 1974: 53).

Despite the simple principle being cemented by the Treaty of Tordesillas, it did not last long in its original form. As the boundary was drawn from 'pole to pole' on a two-dimensional map, it did not take into account that the earth is round. What happened if Spain sailed west and continued? The primary motive for exploration was trade – especially in spices, and the most valuable locus was the island group of the Moluccas in contemporary Indonesia. The Portuguese reached these islands in 1515 by sailing east, which was only six years before a Spanish fleet reached them by sailing west (Destombes 1955: 66). When Magellan's expedition reached the Moluccas in 1521, it disrupted the settlement of the Tordesillas Treaty because both countries could now claim the right to the island under the settlement of this Treaty. It therefore became necessary to decide where the demarcation line would run on 'the other side of the earth', and thus make clear whether the islands were within the Spanish or the Portuguese sphere. To reach an agreement, negotiations commenced between representatives of the two governments during the 1520s.

These negotiations introduced the globe as a tool for international politics. Two delegations were launched from the respective sides, and they presented a large collection of maps and globes in order to support their claim to the disputed islands. Prolific cartographers such as Diego Ribeiro, who had been greatly involved in planning Magellan's journey (Brotton 1997: 133), and the Reinel brothers on the Portuguese side featured among the delegates. One of the key disputes concerned the size of the earth as the length of the Pacific would decide whether the islands would fall on one side or the other. Despite their collective expertise, it was not possible for the cartographers to decide 'scientifically' where the demarcation line should run. The agreement that was eventually reached by the Treaty of Saragossa – ratified in 1529 – probably reflected that the westward journey to the Moluccas was untenable for the Castilian crown as the journey was too long and arduous to be profitable (Brotton 1997; Parry 2000: 161). With this settlement, Charles V gave up his claim to the islands in return for compensation that amounted to 350,000 gold ducats. The Spanish throne also negotiated a clause that allowed the settlement to be renegotiated if new geographical evidence should occur that would support the Spanish claim (Brotton 1997: 136). The agreement was tied up in a map, which had to be recognized by both sides, and '[t]his chart shall also designate the spot in which the said vassals of the said Emperor and King of Castile shall situate and locate Molucca, which during the time of this contract shall be regarded as situated in such place' (quoted by Brotton 1997: 137).

The fact that globes were introduced to represent spatial knowledge on which to base the negotiations underlines how a global view of the world became a reference for politics between European states and,

further, how geographical knowledge was becoming of prime significance to politics. The determination of longitude was no longer only
a matter of navigation but rather a matter of solving political disputes
over land. Interestingly, Portuguese naval charts started to show scales
of latitude from circa 1500, but scales of longitude did not appear until
the dispute arose over the Moluccas 20 years later (Destombes 1955:
76). The question that now poses itself is whether, and how, this new
carto-global politics emerging during the negotiations between the two
Iberian powers became universalized – or made global itself.

The cosmographic globe

A central remarkable feature of the new cartography growing in Europe
is that according to its own principles it becomes a universal tool producing universal knowledge separated from social symbolic significance.
In practice, of course, this cartography was spatially particular, claiming
to produce universal knowledge about space, but according to its own
principles it was universal and was not constrained by anything but the
rules of geometry and the problems of assembling and coordinating data.
Where the Christian *mappaemundi*, as discussed in the previous chapter,
has obvious links with a particular culture, the modern map has less
visible links or origins, precisely because it is emptied of explicitly symbolic value content. The same could be said about contemporaneous
Aztec and Chinese cartographic traditions which centred the map on
the symbolic 'capital' within the world that is represented. In contrast,
scientific cartography has no pre-defined centre; the development and
usage of the coordinate system which meant that 'new places could be
fitted in as their coordinates became available without 'stretching' or
extending the map [and therefore] the Ptolemaic frame could theoretically accommodate discoveries worldwide' (Woodward 2007a: 13). This
represented a major shift and was expressed in the another feature,
orthogonality, which means that the map is seen, not from a single perspective, but all places on the map are represented as seen straight from
above; that is, an infinitude of single point perspectives (Woodward
2007a: 15–6). It is thus important to emphasise that scientific cartography not only claimed to be a universal science of space it was also
applicable everywhere because of its geometrical rationale.

The image of the globe was not an invention of the Renaissance,
but the 'practical use' to which it was put represented a significant
transformation. In preceding periods, the globe had been adopted to
signal universality, and as a symbol of power. In the first century BC, the

globe was adopted by Rome to signify empire and, with the addition of a cross, was later adopted by Frankish and German emperors (Cosgrove 2003: 11). By the twelfth century the study of an ordered creation was put on the curriculum at the new universities being founded in Europe, and the global map (the *mappaemundi*) conveyed a synthesis of Aristotelian natural philosophy, biblical authority and a growing portion of spatial data brought back to Europe by crusaders and other travellers (Cosgrove 2003: 69–71).

Cosmography was part of a university tradition that covered a wide range of areas, including geometry, astronomy, cosmology and so forth (Sandman 2007: 1107). The word 'cosmos' comes from Greek and signifies order so Cosmography was about bringing order to the world. In its task, cosmography was future oriented as it aimed to complete an unfinished image of the globe (Lestringant 1994: 3) in contrast to medieval cartography which was more concerned with the historical past. The drawing of the world during the Renaissance represented thus a preoccupation with establishing a (new) spatial reality. In a sense, cosmography appears as a transition stage between knowledge based on 'pure practical' experience and the more systematic science that was to follow (Lestringant 1994: 131). The scheme which was put forward by the Ptolemaic geography was one that played into a notion of a singular world divided into political particulars. It not only involved the use of a geometrical perspective that came to inform the representation of space, it also involved a distinctive approach between cosmography, which was concerned with the entire known world on the one hand, and chorography which was concerned with particular places (Mignolo 1995: 281). Within the same epistemic framework, then, we get two levels of analysis: the globe as the entirety on the one hand and, on the other hand, the concern with the mapping of particular places.[3]

Martin Waldseemüller's *Cosmographiae Introductio* from 1507, for example, made a break away from the tricontinental world image of medieval geography by adding a fourth continent to the old world (Cosgrove 2003: 97). He named it after Amerigo Vespucci who was thus given the honour of having discovered the 'new world'. Later, Waldseemüller realized his mistake, but by then it was too late and the name America had been established for good (Thrower 1999: 71). Waldseemüller's world maps became hugely popular and influential as they represented the new perspective on universality now that the *oikumene* was being enlarged and coincided with the terrestrial sphere (Lestringant 1994: 24). Consequently, cosmographical studies established the mapped globe within a greater celestial scheme as the universal sphere for human activity. In the words of Lestringant, the 'hypotheses of cosmography supposed a full, global world with no other limits than the celestial orb that, projected onto it, formed its poles, regions and zones' (Lestringant 1994: 12). Defining the framework in terms of the globe, the image was gradually improved and 'completed' with the data collected by navigators and travellers which was combined with the classic textual inheritance.

Under the cosmographic scheme, the global image had to be constantly updated as new data rendered old representations untenable. In establishing the new reality of space, experience became the predominant authority as cosmography brought together the practical experience of the sailor, or the navigator, with theoretical knowledge of geometry (Lestringant 1994: 104–5). In 1464, Nicolas Cusanus described the cosmographer as a creator. He described the cosmographer 'as a man positioned in a city with five gates, representing the five senses. Messengers bring him information about the world using these senses, and he records the information in order to have a complete record of the external world [...]. When he has received all the information from the messengers, he 'compiles it into a well-ordered and proportionally measured map lest it be lost" (Woodward 2007a: 17–18). After having shut the gates and made the map to preserve (store) data, Nicolas concludes that 'in so far as he is a cosmographer, he is the creator of the world'; just as God the creator existed prior to the world, so the cosmographer existed prior to the map (Woodward 2007a: 18). Just how central cosmography, or geography, was considered in relation to general learning is illustrated well by the rather immodest opening by the Florentine scholar Francesco Berlinghieri in his translation of Ptolemy's Geography:

> How many [disciplines] are affected by the delay of this great work, which takes into full view the whole earth. It feeds not only military art but also philosophy, scripture, history, and poetry. The sweet life of agriculture, medicine, and art that animates the love of nature in the human breast. In sum, no greater need have our faculties than knowledge of the earth.
>
> (quoted in Cosgrove 2003: 108)

The cosmographers, then, gathered and reconciled a large amount of disparate geographical descriptions within an overall theoretical framework and, at the same time, navigated between various political and intellectual interests. The establishment of a new reality of space had, to a large extent, become a practical exercise in which the cosmographers' studies – echoing Lestringant's description of André Thevet's 'atelier' – became centres of accumulation albeit, primitive ones. A 'centre of accumulation' is a term introduced by Latour, as discussed in chapter 3, to describe the place where scientific knowledge production 'starts' and takes place. These centres can become centres of accumulation if all connections conspire towards the same goal: 'a cycle of accumulation that allows a point to become a *centre* by acting at a distance on many other points' (Latour 1987: 222). The centre performs the role of a 'hub' in which disparate traits of knowledge are assembled and ordered into a whole and from where a new expedition can be launched.

In effect these workshops worked as laboratories making the world knowable and navigable. They constituted hubs in networks of navigators, surveyors, cartographers, printers, etc. which were all necessary for geometric cartography to develop. 'Cartography is one network cumulating traces in a few centres which by themselves are [...] local' (Latour 1987: 229). Assembling knowledge in this way is all about the possibility of *coming back*. In the words of Norman Thrower, 'a place is not really discovered until it has been mapped so that it can be reached again' (Thrower 1999: 64). Discovery, then, is not primarily about finding new lands but about bringing them 'back home' in terms of knowledge that will allow a return to the place in question (Latour 1987: 217). Without the ability to return, the 'discovery' is next to pointless for the patrons at home. This enables people who have the accumulated knowledge to be able to send expeditions out to places, of which they have no personal experience, and be reasonably confident of the expeditions arriving at these places and returning home. Furthermore, this spatial knowledge would emancipate 'the expedition' from dependence on local guides on whom they would otherwise have to rely in order to get to their destination.

The key issue concerning the ability to come back is that 'action at distance' becomes feasible. This means that those in the centre are able to plan, and act at a distance, on locations where they have not necessarily been and from where people are not necessarily able to reach the centre. To give examples, a king would be able to capture and plan travels in the entire kingdom by looking at a sheet of paper on his wall; the Dutch East India Company would be able to plan the route of fleets in their map room. And it was not only possible to act at distance in known, that is, mapped, places; the abstract Ptolemaic representation of space allowed the issuing of instruction to go into unknown, yet, pre-defined spaces. In 1603, for example, the French King Henry granted royal letters to a Pierre de Guast appointing him to Lieutenant-General to represent the king 'in the countries, territories, confines and coasts of La Cadie [Nova Scotia], commencing from the fortieth degree unto the forty sixth degree; and within the said limits or any part of them, as far and to such distance inland as may be possible, to establish and extend and make known our name, power and authority [...]' (quoted in Keller et al. 1967: 111). This instruction contains a mix of abstract and practical demarcation. The starting point is defined by abstract coordinates and the scope of extending the king's authority into the foreign lands is made a matter of practical feasibility. And 'acting at distance', Latour argues, requires that somehow places are brought back to the centre. This is done by: making them *mobile*; keeping them *stable*; and making them *combinable* (Latour 1987: 223). Captured by scientific cartography other worlds were rendered mobile, taken 'home', combined with other worlds, and hence an accumulation of spatial knowledge was able to take place. This was done by the sixteenth-century European cosmographers and, even though it was done in a less sophisticated and controlled manner than by the eighteenth-century scientific enterprises, their studios can be considered 'centres of calculation' in which knowledge is accumulated.

The French cosmographer, André Thevet, is credited with making the most ambitious attempt to unify expanding oceanic knowledge into a single image of the terraqueous globe of his time (Cosgrove 2003: 99). The centre of calculation was Paris where he was working as the royal cosmographer to the last Valois king, Henri III of France. His work was based on a literal reading of Ptolemy, which he combined with a large amount of chorographic accounts in order to produce a complete picture of the world. Working under the auspices of the French king, he tied his scientific enterprise to the state, and his aim was, to quote Lestringant, to 'transform the intellectual and symbolic possession of the world into a military conquest of it' (Lestringant 1994: 16). Nevertheless, as his cosmography was based on an unsystematic *bricolage* of sporadic knowledge and various ancient sources neither he, nor the state, controlled the input of knowledge and the end product was therefore less systematic than later projects; yet still the traces of the standardized processes of knowledge production are visible. As an example of cosmographic enterprise, this example illustrate how the belief that space and time exist independently as a frame of reference *inside* which events occur 'makes it impossible to understand how different spaces and different times may be produced *inside the networks* built to mobilise, cumulate and recombine the world' (Latour 1987: 228, italics original). In other words, the events we discuss do not take place inside space; on the contrary, space was generated inside the observatory, or the atelier, of the cosmographer.

Assembling the globe

While it is a common popular belief that Columbus' voyages proved that the world was a globe, he simply reached a continent that was (maybe) unknown in his world.[4] The empirical proof that the world was indeed a global sphere was not provided until the circumnavigation of Magellan's fleet during the years 1519–22, yet even prior to that, only few disputed that the world was spherical (Russell 1991).[5] The disputes that Columbus faced prior to his first voyage concerned the size rather than the shape of the earth. Curiously, he was in fact wrong by underestimating the size of the earth by ca. twenty percent in the face of critical arguments bearing on a more accurate calculation of the size. It was, thus, Columbus' good fortune that he – unknowingly – encountered an unexpected continent to cut his journey short.

Columbus' geographical ideas regarding issues such as the expected length of the voyage and the location of the Asian countries are reflected on the oldest surviving European globe completed by Martin Behaim and his team in 1492. Both drew on similar geographical sources (Harley et al. 1990: 53) and Behaim was familiar with the latest Portuguese travels along the African coast. He thus presented current knowledge, and it is clear that Columbus did not set out to navigate a blank or unknown space. On the contrary, if Harley is correct in his argument about converging knowledge between Behaim and Columbus, the latter would have had a fairly clear idea of when he would encounter Japan, or Cipangu, as described in Marco Polo's travels (Harley et al. 1990: 53). Behaim's globe was commissioned in 1490 and, while being completed simultaneously with Columbus' voyage, it illustrates that there was a clear global cartographic image in place prior to the globalization of navigation, trade and conquest.

Even though, it was not immediately recognized that an unknown continent had been found, the encounter with the Amer-Indians in itself shattered the established division of the world into three parts with their respective races. However, it was not only Christian cosmological conventions that were challenged; the recently (re)discovered authority of Ptolemy would also have to be amended in order to make space for a continent that no template had previously allowed for. Direct experience and observation thus became the sources of assembling a new knowledge of the world. The requirement to keep up-to-date with continuous expanding geographical horizons in the Americas was a central challenge. As discussed, the ability to return was key, and in order to do that, the Spanish state set up large scale institutional systems to create a more controlled and comprehensive attempt to establish a fully global image of space following the years after the Treaty of Tordesillas; an exercise described by Cosgrove as 'oceanic globalism' which 'altered both Europe's imperial vision and its constructions of humanity' (2003: 80). This was a matter of filling in the unknowns and correcting the content of the abstract global image. Not surprisingly, mapping initially focused on coastline mappings, thus drawing the contours of the discoveries. Subsequently, more orchestrated efforts would be made to 'fill-in' the contours by establishing topographical knowledge of the claimed territories.

Both the Spanish and the Portuguese authorities established institutions to keep track of the new spatial knowledge very soon after the settlement of Tordesillas. As the globe had been politicized, it was urgent for both sides to obtain as accurate an image as possible. In response they both established 'master maps' which assembled all the disparate pieces of information that had been obtained, primarily, from returning captains. The *Padreo Real* and the *Padron Real* were, respectively, the master charts of the Casa de Mina in Lisbon and the Casa de Contratación in Seville. Through these, the two crowns sought 'to keep knowledge of new discoveries within the control of the state and to ensure the standardization of knowledge, so that errors and inconsistencies among charts could be eliminated and they could be revised and updated as new discoveries were made' (Turnbull 1996: 7). Hence, it was an attempt by the state to create a centralized knowledge framework which would allow pilots, by their own means, to return to any location which had been discovered, and to ensure that this framework was regularly updated.

As such the *Padron Real* served a dual purpose; by providing a single continuously updated standard it was supposed to provide improved and standardized maps to the pilots in Spanish service, and, at the same time, it supported the territorial claims of the Spanish crown particularly with reference to the treaties of Tordesillas and Saragossa (Sandman 2007). It consisted of a large world map and of a book containing descriptions of specific routes (Lamb 1974: 57). It was managed by the Casa de Contratación (House of Trade), which was established in 1503 to control Spanish overseas trade and colonization. In effect it worked as a department of government responsible for, among other things, commerce, a school of navigation, and a hydrographic bureau (Haring 1964: 32). The latter was established in 1508 and the person giving name to America, Amerigo Vespucci, was the first 'pilot major' responsible for the *Padron Real*. Partly to remedy the problem of poor duplicate maps being sold in the Spanish ports, Amerigo Vespucci was instructed to create an official master map to minimize errors (Haring 1964: 306). Turnbull quotes Vespucci's instruction at length:

> We command that a Padron General be made and, so that it should be more accurate, we command our officials of the Casa de la Contratación that they assemble all our pilots, the most skilled captains at the time, and that the said Amerigo Vespucci, our pilot major being present, a padron of all the lands and islands of the Indies hither to discovered and belonging to our kingdoms and seignories be drawn up and made ... when they find new lands or islands or shoals or new harbours or anything that should be recorded in the said padron real on their return to Castile they to report to you the said pilot major of the Casa de la Contratación so that all shall be registered in the proper place in the padron real, in order that navigators be better advised and cautious.
>
> (1996: 11)

In that respect the *Padron* worked as a mediator between the pilots and the spatial locations they were exploring. While different initiatives were tried and tested during the century, information from the pilots remained the most important source for improving and developing the *Padron*. From 1527 the pilots were ordered to keep detailed records of their journeys keeping track of 'the path which they followed every day and on which rhumbs, and which lands or islands or bays they reached, and how far away they were, and how the coasts ran, and what ports or rivers or capes there were in these places, and in what distance and latitude they lay' (official instruction quoted by Sandman 2007: 1101). These reports were to be submitted to the Casa on their return in order to improve the *Padron*, and as such, the Casa would serve as a centre of calculation assembling the regular inflow of new spatial data from the returning pilots into a gradually completing global map which, in turn, would serve the pilots on their travels and the king in his territorial claims.

While in theory this sounds straightforward, in practice there were several difficulties along the way. The one concern spatial data; how to decide which were reliable when pilots' observations differed, and how to decide when a revision of the *Padron* was needed? There was a significant mistrust between the practitioners in the field and the theorists in the Casa, between cosmographers with diverging interests, especially between those that had made a lucrative trade in charts and other navigational instruments and those primarily occupied with the *Padron*.

Another significant obstacle was created by mismatching cartographic epistemes. The *Padron* combined the portolan tradition of Mediterranean navigation with the Ptolemaic scheme. The portolan tradition was not based on the grid system of latitude and longitude but organized a complex web of 'rhumblines' specifying the compass bearings between any two places on the map (Sandman 2007: 1096). The *portolans*, presumably, worked well as a specific mode of producing knowledge about a limited area, but the rhumbline structure were not adequate as a framework for coordinating an open, or unfinished cartographic global space. This is not the place to go into technical details but basically it is not possible to reconcile straight compass bearings on a navigational chart without distorting distances, and for oceanic travel the *portolan* style proved inadequate. It was, therefore, a complex task to reconcile fragmented pieces of knowledge into the general framework and slowly the status of the *Padron* would erode until, by the 1560s, it 'slipped into disuse' (Turnbull 1996: 14).

Despite these difficulties the *Padron* represented one of the most significant attempts to establish a cartographic institutional structure aimed to establish the global world cartographically. As a centre of calculation, the Case produced a unified space which they were able to maintain and update, even if they did not complete it, and thus produce an expanding space that was navigable and to which it was possible to return. And even if not entirely successful, the project still represented an unprecedented ability at the time to equip navigators with uniform maps and, in turn, enable planning and coordination of large fleets over long distances and long time spans.

Mapping new worlds

Up until the mid-sixteenth century, global mapping done by Europeans had been mostly concerned with the mapping of coastlines as the immediate need that was driving cartographic innovations came from navigation and the desire to produce a navigable global space. In addition, the treaties between the Iberian powers generated a preoccupation with longitude even though this was notoriously difficult to determine until the eighteenth century. However, as Castile faced a challenge of governing the new possessions after the conquest of New Spain (Mexico, conquered around 1519–21 by forces lead by Hernán Cortés) and Peru (conquered just over 10 years later by Pizarro's forces (Parry 1990: 83–7)), the need for spatial knowledge increased. Governing these new possessions meant governing a space beyond the reach of the monarch (Mundy 1996: 8); Isabel and Ferdinand, for example, had travelled throughout their realm during their time in power in order to maintain a presence there. The task of government required the throne to obtain spatial knowledge, which could be used to act independently of indigenous knowledge, as well as, local governors, and to provide the basis for a more efficient government of 'New Spain' from Old Spain. The distance to the colonies increased the problem of loyalty, and the Crown feared that the new elite, gaining their riches and prestige in the Indies, would not remain loyal to the crown (Parry 1990: 88).

As such, it was necessary to make the colonies present at the centre of the empire. It was from the 1550s that the resources of the new Spanish territories started to reach Spain on a large scale (Goodman 1988: 173; Elliott 2000: 29), and it was in the 1550s that the Spanish crown launched a series of attempts to map their new possessions by questionnaire. Especially from 1569, more concerted efforts were made to bring descriptions of the Indies back to Spain. This was done under instructions from the Council of the Indies, which had been established with the responsibility of overseeing the government of the transatlantic empire and it did so 'in superbly bureaucratic style, various instructions and questionnaires poured from the Council's desks to find their way to the remotest New World jurisdictions' (Edwards 1969: 17). In an early project to map the American possessions, the appointed cosmographer Santa Cruz planned to establish a geometric projection of the 'New World' based on the grid system of latitude and longitude (Mundy 1996: 12).

The project was incomplete when Cruz died in 1567 (Mundy 1996: 17) but was taken up again when reorganization of the Council of the Indies created the position of *cosmógrafo-cronista mayor*; a title that emphasizes the link between the textual recording of history and the cartographic depiction of geography, and how these were combined in order to grasp specific localities. López de Velasco was appointed chronicler and cosmographer major and came to coordinate what is known as the *Relaciones Geográficas* in 1577 (Cline 1964: 346). Based on a large number of questionnaires containing 50 questions, local officials and governors were to describe the geography, history, mythology and so forth of the areas which they governed. Furthermore, they were asked to make very detailed measurements of lunar eclipses in order to determine the longitude of the location based on prescriptions by Velasco's predecessor, Santa Cruz, who developed new ways of measuring longitude based on observations of lunar eclipses. On this basis, the *Relaciones Geográfica* questionnaires included a very detailed instruction to the respondents of how and when to monitor a lunar eclipse (Mundy 1996: 18). In addition they asked for maps, or drawings, of the areas in order to provide some kind of topographical mapping. If this ambitious project had been successful, colonial officials would have supplied 'López de Velasco with accurate topographical and coastal maps, as well as precise lunar observations [... which would have provided ...] ample information to make both geographic and chorographic maps of this crown jewel of Spain's colonial empire, thus making this part of the New World visible to its absent king' (Mundy 1996: 22). This would have provided the state with a much needed tool of governance; in the instruction to the governors of the *Relaciones Geográfica*, King Philip II wrote: 'so that the Council can attend to their good government, it has seemed a proper thing to decree that a general description be made of the whole condition of our Indies, islands, and their provinces, the most accurate and certain possible' (Cline 1964: 363).

Yet, the *Relaciones* did not represent a success despite the grand plan. Whereas the Padron Real had suffered from an insufficient spatial framework at home, in the 'centre of calculation', the *Relaciones* were based on a strictly Euclidian framework organized around the grid work of longitude and latitude and was able to contain disparate spatial data on the condition that they were produced in the right way, that is, in a fashion that made them combinable. However, the respondents of the questionnaires did not render the world mobile in such a sense. Velasco had written the instructions on the lunar eclipse observations in a fashion so that a lay person would be able to return 'scientific spatial data' required by the 'experts' back home for ordering the territories according to latitude and longitude. But the almost two hundred replies were all very different – incompatible both with each other and with the rational framework prepared 'back home' by Velasco – and hence did not provide combinable spatial data to fit into the framework. And although some were fairly accurate, such as the observations made for Mexico City, which located it within 35 kilometres of its current coordinates, the project did not succeed in establishing a precise network of positions by celestial observation (Edwards 1969: 22–7). Equally disappointing, from the point of view of Velasco, were the topographical maps which the respondents had been requested to produce. The majority of these had been painted by Native Americans who were drawing heavily on their own map-making traditions with their associated iconography (Mundy 1996). The result appeared as an intricate hybrid of new cartographic conventions and the indigenous traditions which were largely incommensurable with the Ptolemaic episteme of map-making into which the maps were supposed to be combined.

Although not being entirely successful, both the *Padron Real* and the *Relaciones Geográfica* stand as significant examples of the transition toward a new epistemic mode of mapping the world. In the case of the *Padron*, the framework receiving spatial data was based on a tradition preceding that of geometric cartography, the portolan, which was based on practical navigation rather than the mathematically informed principles which were required to accommodate the increasing world of the Europeans. On the other hand, the *Relaciones* represented a significant encounter between two cultures whose modes of establishing a spatial reality were radically different. As a consequence, the spatial data that were returned to the centre were not *combinable* and would therefore not contribute to a single rationalized global map. This encounter, however, between the different map traditions highlights the power of map-making and speaks to a wider concern with language, identity and cultural difference which have occupied postcolonial studies in recent years. This work has highlighted how certain knowledge structures have produced an inherently unequal world based on particular standards of European knowledge production (most prominently Said 1995).

Several postcolonial scholars have emphasized the centrality of space and the conceptual appropriation of space in the processes by which Europeans expanded their world to incorporate an increasing proportion of the globe. Mignolo, for example, stresses that '[o]ne of the particularities in colonization processes is that people from quite different cultures come face to face and fight to preserve or appropriate territories, both as possessions and conceptualizations of space' (1995: 333). The essential tension arises as a consequence of Europeans entering a world unknown to them and their need, then, to establish knowledge about it based on their own particular principles often dramatically different from the principles of those who inhabit the area. Simply put: '[t]he *terra incognitae* of the white intruders were the *terra cognitae* of the natives' (Lewis 1998: 2). Emphasizing how cartography served the need of the Spanish state to produce the colonies as intelligible units in order to rule them, Raymond Craib argues that 'maps were fundamental to creating the object to be possessed' (2000: 17). This implies that the mapping of, for example, New Spain was not a process of uncovering a pre-existing reality, but rather, of creating one. This theme was raised in Edmundo O'Gorman's classic *The Invention of America*, in which he traces how America 'developed from a complex, living process of exploration and interpretation which ended by endowing the newly-found lands with a proper and peculiar meaning of their own, the meaning of being the "fourth part" of the world' – first described as a separate continent by Martin Waldseemüller in *Cosmographiae Introductio* published 1507 – rather than a sudden discovery by the man Columbus (O'Gorman 1961: 124).

Where O'Gorman celebrates this development as a liberation from the old doctrines of a narrow *Orbis Terrarum* – a world island surrounded by sea demarcating the limited oikumene of humanity, postcolonial scholarship has emphasized how the power of naming favours one and dispossesses the other side in a cultural encounter like the one taking place in the Spanish attempt to take the new lands into possession. In her celebrated study of the *Relaciones*, Barbara Mundy describes how European naming gradually deprived most of the indigenous people of the means to represent their communities. This was because they could no longer recognize place names as there was 'a yawning gulf' between alphabetic and logographic writing in the images (Mundy 1996: 164–9). However, although the hybrid responses to the *Relaciones* survey were incompatible with the grid framework of European spatial knowledge, they proved that the immediate encounter between the two cultures was not incommensurable. Most of the maps contained a mix of the Ptolemaic style of the Spaniards and the cartographic mode of the natives building on a social projection, where societal hierarchies and relations provide the ordering principle of the map (Mundy 1996: xiv–xvi). In this respect, the native Amer-Indians played a significant part in producing the imperial reality. This was not an even-handed process, however, but one in which the epistemic rules were set by the Spanish. While, indigenous cartography eventually had to give way to the European style of mapping in order to represent and secure property and to represent a territorial space of belonging (Craib 2000: 23–7), it significantly influenced the responses to the *Relaciones*. In that, the indigenous population retained some power of authorship even if they had to succumb to the epistemic power of Ptolemaic cartography. As such, the epistemic encounters of the *Padron* and the *Relaciones* are instructive of the challenges facing the Crown of Castile in expanding their activities towards a global scope, and also of the consequences of the epistemic encounter between two modes of mapping. While the indigenous populations gradually had to adopt the epistemic mode of cartography set by the Spanish, European cartography on the other hand relied on local spatial knowledge up until the point where it became possible to deploy, from the centres of calculation in Europe, surveyors who would carry out their own surveys of the colonies governed, and thus achieve independence from native knowledge.

A decisive feature of the universalization of the geometric map was its, in principle, lack of immanent centre built into its epistemic prescriptions. Historically, most cultures have located themselves at the centre of the map. The medieval *mappaemundi* were centred on Jerusalem occupying a prime position in the cosmological imagination at the time. Chinese cartography, which utilized a grid system as the basis of cartographic representations contemporaneously with Ptolemy in Europe (second century AD), remained Sinocentric (Thrower 1999: 30–2). Of course, most European world maps put Europe at the centre of the map, but this was not written into the instructions of 'how to make a map'. Mignolo describes this when he distinguishes between cartography based on 'a geometric rationalization', as opposed to 'an ethnic rationalization', where the former is not based on an ethno-symbolically defined centre of the world (1995: 257–8). He illustrates the point with a tale about the Jesuit missionary Matteo Ricci, who in the 1580s, was paid a visit by Chinese Mandarins who were astonished to see a world map decorating Ricci's walls. Mignolo presumes that this was the Ortelius world map from 1570 which, to the surprise of the Mandarins, showed China as only occupying a limited part of the world and being pushed all the way to the eastern edge of the map. In order to accommodate the world view of the Chinese officials, Ricci changed the layout of the map so that Europe and America were pushed to the sides and China was placed in the centre (Mignolo 1995: 219–20). This did not, of course, change the size of China on the map but it recentred China in the middle of the map and thus satisfied the Chinese notion of 'Middle Kingdom'. Another illustrative example can be found in Inca mapping which divided the Inca territory into districts according to a grid (*ceques*) which, like spokes, radiated from the capital Cuzco following the surrounding mountain landscape. While this grid structure also built on an abstraction of space, the centre was not flexible and so, it could not move. It was a system specifically developed for the site of Cuzco (Turnbull 2000: 27–31).

These examples illustrate an ability of the geometric map to accommodate various 'symbolic centres', which would not have been possible with contemporary Chinese or Inca maps because such a cutting exercise would have clashed with the meaning of the map. This ability to accommodate a variety of centres within the same mode of cartography contributes to the power – and the opacity of this power – of the geometric map; because while submission to the geometric discourse remains a requirement it allows, at the same time, multiple centres to exist. In other words, it becomes possible for different groups to claim authorship and write themselves into the centre of the map. It is the distinction between socio-symbolic content and mathematical projection of space that makes this possible and, if we return to the argument concerning Wolf's imaginary journey, we start to see the contours of a wider argument about global space as being a natural spatial arena for social practice. The Ptolemaic projection produces space-equalizing and area-fixing properties of the graticule, and this frictionless extension privileges no specific point (Cosgrove 2003: 105).

Publishing the world as a stage

In 1599, the Globe Theatre was first built in London emphasizing the linkage between globe, stage and actors. In Renaissance cosmography, the sphere was both a metaphor and a basic unit of order seen as 'the embodiment of a divine geometry' (Short 2004: 34), and the theatre was a common metaphor for the universe (Frye and Denham 1990: 204). In Shakespeare's *As You Like It* performed at the Globe in 1600, the character Jaques declares: 'All the world's a stage. And all the men and women merely players' (Jacquot 1957: 345). This was 30 years after Ortelius had published his *Theatrum Orbis Terrarum* – theatre of the global world – which is known as the first published atlas of the world in Europe.[6] The term theatre reflected a European obsession with microcosms (Akerman 1995: 139), and just like the theatre was composed of a stage and corresponding persons acting on this stage, so the world and the individual were seen as the two main building blocks of the cosmographical world. In cartography this was reflected in a widespread tendency to combine mapping of the globe with the mapping of the human body. In Martin Borrhaus' *Elementale Cosmographicum* (1539), the world is divided into 5 climate zones which correspond to the five fingers of a hand (Short 2004: 44); in Peter Apian's *La Cosmographie* (1544), geography as the description of the earth, and chorography as the description of particular places, is compared to a portrait of the head and the eyes or ears respectively (Lestringant 1994: Pl.2); and Henri Hondius and Jan Jansson wrote in the preface of their five-volume atlas in 1646 (translated and quoted by Cosgrove 2003: 157):

All visible creatures made by God are comprised by these two here, Man and the World. The former has been made lord of the Universe, the latter the seat of his empire. The former is the guest and inhabitant of the world, the latter the most magnificent and spacious house for such a great guest. In Man we recognize the image of the excellent Artisan who created him, and in the world, the image of Man.

Of course this was part of a wider development within the sciences, where tellingly, Copernicus' seminal treatise on astronomy *De Revolutionibus* was published the same year as Vesalius' foundational text in anatomy *De Fabrica* in 1543 (Gribbin 2002: xvii). These texts indicate a growing concern with the human body as well as with the human environment. This was also a time that, in the words of Helgerson, saw 'a self-consciously new literature and a still more self-consciously new geography', which mutually encouraged and confirmed each other (1998: 1). These developments emphasized two 'levels of analysis', so to speak: that of the stage, the Globe, and that of the actor, man, and subsequently the state. Publications like those just mentioned disseminated a novel picture of the world to the literate Europeans and cemented the dramatic change of the world from the medieval *mappaemundi*.

As a consequence of the emergence of America, European cosmography faced a challenge of recreating the spherical order of Ptolemy's geography depicted on Behaim's globe. Ptolemy's map only covered 180 degrees of latitude and Behaim's 1492 globe showed the three continents of the 'old world' as covering the entire globe. The cosmographers of the European renaissance were, therefore, faced with the necessity of making spatial data mobile and stable in order to bring the new worlds back in a form that was compatible with the framework at the centre. Within the first decade of the sixteenth century, the world map had been expanded to cover 360 degrees and thus completing the sphere, and America had been established as a fourth continent (Woodward 1991: 86). As discussed, the treaties and Magellan's circumnavigation rendered the single globe a political space, and this dramatically increased the demand for globes. Nuremberg became a centre for globe-making not only because of Behaim but also Johann Schöner who became the first professional globe maker and produced several globes between 1515 and 1535 (Thrower 1999: 75). Antwerp was another important centre with Gemma Frisuis famously producing globes in the 1530s, some of which were produced in collaboration with his now famous student, Mercator (Short 2004: 28).

These globe projects were not under strict state auspices, as was the case of the *Padron* and the *Relaciones*, but were creations of a private map industry which, however, was still partly under the patronage of the state.[7] They represent a growing industry of publishers and surveyors exchanging and compiling spatial data within a European network of cartographers. For example, one of the most prominent cartographers of the sixteenth century, Gerard Mercator, who was based in the Low Countries, was commissioned by a group of merchants to produce an accurate map of Flanders. It was published in 1540 and provided such a significant improvement compared to previous maps that he was subsequently commissioned to produce a terrestrial globe for Charles V. He was so pleased with the result that he then commissioned Mercator to produce a set of surveyor's instruments for him to bring along on his military campaigns. Subsequently, Mercator became a member of the royal household of Charles V (Brown 1949: 157–9, 170).

In the attempts by the Castilian state to establish a global space their efforts had been divided according to two 'traditions', oceanic mapping focusing on the contours of coastlines and navigational purposes on the one hand, and more topographical style mapping focusing on filling in these contours on the other. In the large cosmographical projects of the mid-to-late sixteenth century these traditions were merged into great global mapping projects doing both. They were following the cosmographical scheme as laid out by Ptolemy of cosmography and chorography – adjusted to a three tier scheme in 1524 by Apian to cosmography, geography, chorography – concerned with describing localities within a single order. In that respect, the different concerns of navigation and surveying were tied together within the same project and, as a result, the map makers would be concerned both with surveying and navigational concerns. And, as such, cosmography developed a specifically technical gaze in order to capture the essence of the world (Short 2004: 40).

It would be difficult to complete the narrative about the powerful impact of European cartographic practices on the world without taking the impact of printing and the ensuing distribution of spatial representation into account (Craib 2000: 22). The relocation of book-making from the monastery to the print workshop had a profound impact on the collection and accumulation of data (Eisenstein 1979: 88). As soon as one edition had been printed, new data could be added to the framework which could be rapidly expanded and developed. As such the printing industry provided a way in which individual publishers could gradually update and complete the novel global map. As a good example of this, Abraham Ortelius not only provided the first atlas of the world to Europe, but at the same time, he made available a truly global scale to the literate population for the first time (Eisenstein 1979: 448). This atlas became incredibly influential, not only within cartography, but also in its wider uses. Charles Vs successor, Philip II, for example, personally owned three copies of this work (Parker 1992: 124) which had sold out within three months of its first issue (Brown 1949: 162). Ortelius had made a living out of the map trade. He travelled and collected the latest maps which he then coloured and resold at book fairs. One of his best customers was the merchant Aegidius Hooftman who, according to Brown, 'bought every chart, he could lay his hands on' in order to find the best routes for his merchandise at sea and at land where it was important to avoid war zones (1949: 161). Ortelius, thus, was very well connected to a network of the best map producers in Europe and it was on the request of Hooftman that he produced his Atlas. Instead of working with a large number of maps, differing in size and scale, Hooftman asked Ortelius for a uniform collection of maps that could be printed. On completing this request, Ortelius – encouraged by his friend Mercator – decided to publish the collection and a commercial and cartographic success was born. By his death in 1598, Ortelius' Atlas had been published in at least 28 editions in 5 languages and it continued to be published in new editions until 1612 (Brown 1949: 164).

Due to the way in which Ortelius had compiled his work, his office effectively worked as a hub for a large network of map-engravers in Europe. From the beginning, it was this network which made possible the publication of the *Theatrum Orbis Terrarum* and each subsequent edition presented maps of new areas and updated maps of areas already covered. Cartographers were very eager to send Ortelius their latest maps and gave advice on how to make improvements. In return, Ortelius made careful dedications to the origins of all the maps he printed. Structure-wise *Theatrum* opened with a world map presenting the entire setting. This was followed by maps of the four continents and then maps of various countries and regions (Ortelius and Skelton 1964). In this way, he followed the two-tier order of cosmography and chorography describing the stage and the actors. Apart from its form and structure, the *Theatrum* set itself apart by representing a turning point away from Ptolemy in that Ortelius did not present his work as an edition or an update to Ptolemy. In a message to the reader, Ortelius notes how the 'limits of ancient knowledge were graphically and textually underlined so that the reader could see "how maimed and imperfect" were ancient world views which comprised 'scarce one quarter of the whole globe now discovered to us' (quoted by Eisenstein 1979: 193). Inheriting the success of the *Theatrum*, Mercator's *Atlas sive cosmographicae meditationes de fabrica mundi et fabricati figura* took over the leading position after its completion in 1595. Where Ortelius had excelled as a publisher, Mercator is accredited with making cartography more scientific. He carried out his own thorough surveys and, in 1569, he published the first map using what became the notorious Mercator projection that became standard for navigation charts and is still used for map production today. Mercator's atlas presented technical improvements when compared to Ptolemy in that, for example, the different map pages showed overlapping areas so that one could trace a continuous route between the pages. This also meant that the maps were much more uniform than Ortelius' maps which originated from different map-makers (Crane 2002: 253; Akerman 1995: 149).

Ortelius' and Mercator's two atlases became a new standard reference instead of Ptolemy. Willem Blaeu, who came to dominate the industry, published his first atlas in 1631 and described it as an appendix to Ortelius' and Mercator's work (Brown 1949: 171). With Blaeu's dominance, Amsterdam became the absolute centre of the European map industry and the family also came to play a dominant role in the expansive movements of the Dutch commercial empire. In 1633, Blaeu was appointed head of the Dutch East India Company's department of hydrography and he also received the title 'Map Maker to the Republic'. His dominance is reflected in the ruling that merchants dealing with India were allowed to send no other maps than those produced by Blaeu on their ships and hence, in effect, Blaeu had monopolized chart making in the Dutch Republics. It was in this intersection between state and corporate patronage and private entrepreneurship as a commercial publisher that Blaeu established his fame as a publisher and produced, what some consider the most beautiful geographical work ever: *Atlas Maior* published in the 1660s (Brown 1949: 168–73). This 12-volume atlas was a compilation of all geographical knowledge assembled in the main centre of European map-making, but, as argued by Cosgrove, it also appears as a 'total artwork, synthesizing, illuminating, and celebrating an imperial mastery of creation [...]. The whole work acts as a totalizing emblem of knowledge, illumination, and global acquisition' (Cosgrove 2003: 158).

Where the Spanish projects to establish a global cartographic space had been centralized and controlled efforts to collect and assemble spatial data, the atlas makers of the Netherlands relied to a larger extent on a growing international network to gather their geographical information. As surveying techniques and instruments were developed and Europeans states became increasingly concerned with mapping their realms, as I will discuss in the following chapter, more detailed maps became available from more and more countries. In addition, the Dutch East India Company, under Blaeu's instructions, issued orders to their pilots to observe and record eclipses and take measure of relevant locations (Brown 1949: 171). The Dutch cartographers thus utilized tactics similar to the ones used by the Castilian state, but also benefited largely from being part of an international network of scholars and publishers. And where the two traditions of oceanic mapping of coastlines were separate from land surveying, the atlases combined the entire surface of the globe in a single volume. It was thus the marriage of the private entrepreneurship of commercial publishers and humanist scholars with states with imperial ambitions, and large companies such as the Dutch Indian Companies, which produced a single global cartographic space.

The atlases not only brought the world back home, they also rendered this world movable and, with later pocket editions, the world could be carried around, studied, comprehended and action planned at one's convenience. To restate the opening quote of this chapter from a satisfied customer of Ortelius: 'You compress the immense structure of land and sea into a narrow space, and have made the earth portable, which a great many people assert to be immovable' (quoted by Brotton 1997: 175). This fitted well with the heliocentric ideas about the universe, first published by Copernicus in 1543 where the globe was no longer a static centre of the universe but a smaller part in a larger whole.[8] At the same time, the atlas provided a representation of the globe as being a single space, navigable by entrepreneurs in the name of colonization and commerce, and it thus provided the global scale for the actors of an emerging European international politics. Rephrasing Chandra Mukerji, atlases like Ortelius' did 'present the world as a singular and knowable entity' (Mukerji 2006: 660) and thus substantially unified the globe as a single space. Instead of being the partly unknown abstract entity of the Treaties, discussed above, the globe was becoming an abstract, but known, space to the Europeans who were then able to spread their world of commerce, civilization and Christianity in their efforts to merge the European world with a global world.

It was no accident that Amsterdam became a leading site in European cartography. Following the revolt of the Netherlands against Habsburg rule in 1566, only the Northern part stayed independent and was thus subject to an inflow of merchants, scholars and other people from the southern provinces, including the significant trading centre of Antwerp (Lesger 2006: 1–13, 258–63). During the truce between Spain and the newly founded Dutch Republic from 1609–21 Amsterdam manifested itself as a significant locus in European global politics. With the Dutch *de facto* independent, and the decline in Spanish power after 1600, they were able to navigate and enter areas hitherto considered to be within Spanish and Portuguese spheres of influence. This new actor on the global stage 'provided a naval screen behind which the English, the French, the Scots and the Danes, without much danger of Spanish interference could build up their colonies [...] down the Atlantic coast' (Parry 2000: 189; see also Elliott 2000: 81–93, 190–200, 241–55). By breaking the Iberian hold of the non-European world, more actors were competing on this stage for access to and control over global space. And hence, the dream of a Spanish worldwide empire – symbolized by the globe – was left behind in the face of interstate competition.

Conclusion

By virtue of making 'representations' of a given world, the cartographer commands a measure of autonomy; and this, in a sense, poses a challenge to any notion of a given, or natural, unity of space. While playing down the causality, or the rationale, behind map-making, this chapter has aimed to illustrate some of the dynamics that served to politicize the globe and, at the same time, unify the globe through cartography. In short, I have aimed to show that a global space preceded and was completed as European interstate and commercial relations 'globalized'. The globe was rendered a political space in the Iberian competition for trade routes to India during the second half of the fifteenth century. The two treaties of Tordesillas and Saragossa conceptually required an abstract global space, which in its spherical form was in place when the treaties were ratified, but its content was lacking. In response European cosmographers gradually completed the global image by making 'other worlds' mobile and create a unified cartographic representation being able to incorporate a global spatial totality. An important dimension of scientific cartography is made visible in the encounter between different map epistemes. Dissolving the symbolic centre of authority, the scientific map can, in principle, be applied anywhere and centre on any location in the world. The last section of the analysis emphasized the publication of the globe as a stage. Private Dutch map-makers and publishers, working under the Hapsburg state or large private companies, completed a mobile representation of the globe as a singular world-stage which provided the natural background against which individual states would act.

Of course, global mapping was not completed by the time of the publication of Blaeu's *Atlas Maior*, but the great atlases of Ortelius, Mercator and Blaeu set a standard for how to view the world. For overseas mapping, Europeans had to adopt local knowledge until they were technically able to coordinate large-scale surveys such as the ones undertaken by the British in India during the first half of the nineteenth century (Edney 1993: 61). In that respect it was a long and arduous process to complete the representation of global space according to the scientific standards inherent in the Ptolemaic episteme. The idea that there was a global space which had to be completed underlies the analyses of the notorious geopolitician Mackinder who famously stated that with the exploration of the North and South Pole 'the book of the pioneers has been closed. [...] Whether we think of the physical, economic, military or political interconnection of things on the surface of the globe, we are now for the first time presented with a closed system' (Mackinder and Pearce 1962: 29). What was new for Mackinder under this condition, writing in 1919, was that there was no longer any elasticity of political expansion. The entire system was known, the stage was locked – and in a fashion echoing claims to globalization – all places were interconnected such that an event occurring in one location would have effects far beyond this location. Hence, the recent war had involved 'every considerable state' (Mackinder and Pearce 1962: 30). And this system, I have argued, is based on Ptolemaic mapping principles and was inaugurated by the Iberian states, the Pope, and renaissance cosmographers from the end of the fifteenth century.

A similar notion of completion can be read out of Bull and Watson's studies on the *Expansion of International Society* in which they state that the evolution of the European system of interstate relations and the expansion of Europe began at the end of the fifteenth century and 'were concluded by the end of the Second World War' (Bull and Watson 1984). However, in their concern with the emergence of an international global system, Bull and Watson focus on 'rules' and 'institutions' that govern the interstate relations. Implicitly, global space is there just waiting to be flooded by European mariners, merchants and missionaries. Yet, according to my argument, it would be flawed to consider space as something simply present and just waiting to bring social unity to the globe. To regard the process of mapping the globe as a victory of unified science in the name of accurately being able to map the world would be misleading. Rather, it should be considered as a social transformation which participates in transforming the nature of the international itself.

The spherical basis of scientific cartography, firstly, provided the global space before this in any sense had been established through social practices. Secondly, it was the geometric cartography which allowed the European cosmographers to bring the rest of the world back to Europe and, hence, enabled the possession through knowledge of, and coordination of action at distance on, the gradual expanding spaces claimed by European states. Furthermore, the transition from an 'ethnic' to a geometric rationalization of space has in several ways conditioned the establishment of a global state-system.

It is because the scientific map projects space as something natural – as opposed to something cultural – that we can maintain the understanding of a unitary global space in which various social groups can interact with each other, rather than considering the spatial dimension part of this social interaction itself. In the polemic essay, *War of the Worlds*, Latour argues that the most significant impact of 'the West' in the history of the world is the 'insertion of a unified nature' behind cultural differences (2002b: 11–12). This makes conflicts over space simply social or cultural affairs while leaving 'the unified space of nature' as an undisputed settlement. In sequence, this points to the historical necessity of dissolving the notion of a unified global space as a natural fact. What needs to be recognized is that the notion of a unified global space has been made possible as a consequence of a specific mode of mapping space which unifies while simultaneously allowing for a socio-political differentiation. Not as a natural fact to be uncovered, but in cartographic practice which in itself is social.

Thus, as long as there remains a unified cartographic production of the globe, the spatial underpinnings for socio-political agency with a global reach are in place. Rather than being antagonistic to a territorial political organization, it is possible that the cartographic establishment of a spatial global reality provided an incentive for state rulers to become authors of their own world, rather than having some Dutch map-maker decide the limits and nature of their territory. Whether this is generally true, it is noteworthy that several European states initiated surveys and mapping projects during the sixteenth century, as we shall see an example of in the next chapter. And at the same time more and more atlas maps delineated political boundaries indicating that atlases became increasingly sensitive to political territoriality (Akerman 1995: 139, 144).

Rather than considering territory a static entity, one needs to keep in mind that it was because the scientific map renders space 'mobile' (and later, the chronometer makes it possible to keep track of time in other places) that European culture was made mobile. And this mobility and these knowledge networks preceded and conditioned the constructions of allegedly static territories. As will be the focus of the following chapter, cartographic practices enabled the formation of centralized territorial states, and yet, on a more general level the dissolution of the absolute centre inherent in 'ethnic' rationalizations of space must be a precondition for thinking multiple sovereigns within a universalized global stage. As space was established as empty and Euclidian, it was possible to have a plurality of formally equal actors on the stage of the world and, with the decline of Spanish Habsburg power around 1600, no European state could likely claim the ambition of ruling a single world empire. It was thus no accident, as Cosgrove informs us, that oceanic discovery coincided with changing spatialities within Europe as absolute territorial sovereignties emerged (2003: 83). On the contrary, it seems that the globe and the territorial state have come hand in hand.

6
The Cartographic Formation of Denmark

Claudius Claussøn Swart[1] was the first 'geometric cartographer' to map the Nordic countries and introduce Greenland into the 'Ptolemaic world map'. He was born in 1388 on the Danish island of Fünen and later travelled to Italy where he was involved in the new cartographic science emerging there. His maps, which were published in the 1420s, are known as the first additions (*tabulae novae*) to Ptolemy's geography, and they became widely circulated in Italy and Germany and provided a standard for the depiction of Scandinavia for a long time (see map 6.1). It is possible that the maps were made following a commission by the Danish king Erik af Pommern to produce a map of the Danish realm, when they met in Venice in 1424 (Nørlund 1943: 12–16; Bjørnbo and Petersen 1904). Swart would thus have been the first Nordic cartographer to produce a general map under royal command.

Swart's story illustrates the structure of a bourgeoning European science of space, where Ptolemy remained the key authority and new knowledge would gradually be added to the existing body of geographical knowledge; its reliance on individuals and the network structure of science, and its association with state power. Swart's story points to how the emerging princes of fifteenth-century Europe showed an increasing interest in employing this new science to mark out their own realms. Whether the King's request was followed or not is less significant than the request itself. During the sixteenth century, European state rulers were increasingly occupied with mapping and measuring space in the attempts to enhance the territorial government and defences. This is to say that rulers sought to become authors of their cartographic space by commanding the authorship position of cartographic practice.

During the period, where geometric cartography were increasingly becoming the subject to sovereign interests ca. 1450–1650, the European

 121

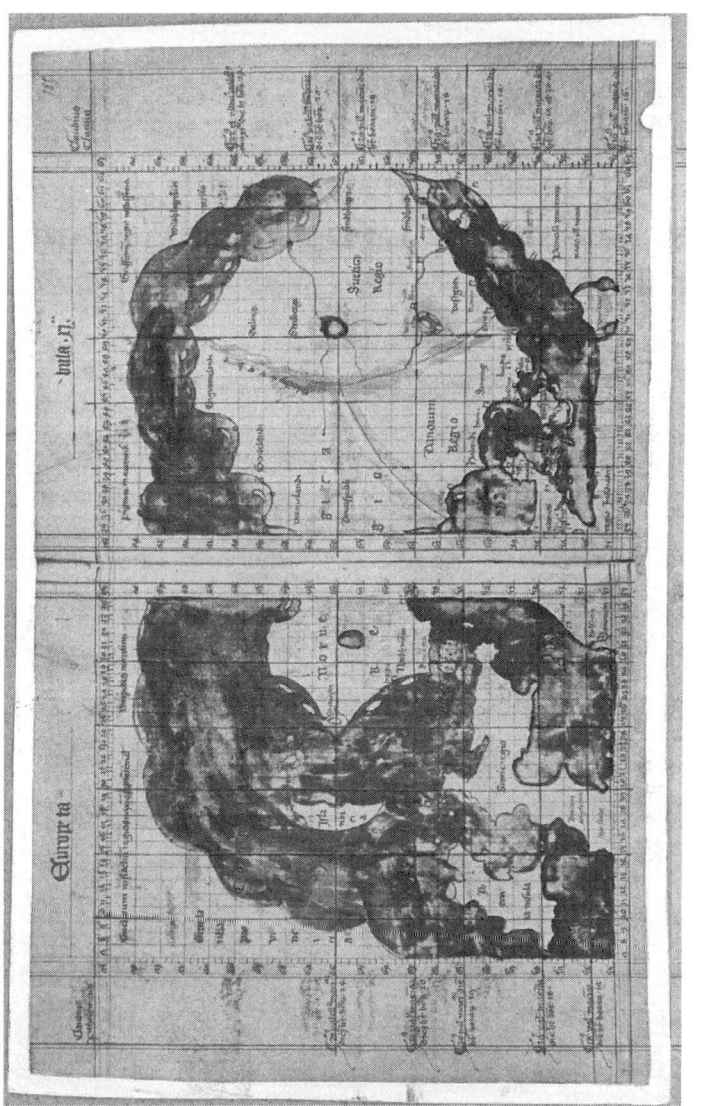

Map 6.1 Swart's map over Scandinavia. Note the grid lines that make it fit into Ptolemy's general framework. Adding Greenland to Ptolemy's map also expanded the world westwards. By courtesy of the Royal Library Copenhagen.

political system changed from containing a number of polity types, some of which were territorial, to a political system increasingly dominated by territorial states claiming sovereignty over their entire realm. This transition is linked with a transformation of the territorial identity of the state to a situation where the state became increasingly identified in terms of its territory. Where pre-cartographic territory was constituted through a web of personal relations between the ruler and subjects, the cartographic transformation enabled the representation of a permanent continuous territory which came to signify the spatial body of the state. It was also during this period that maps were increasingly adopted by state bureaucrats. While almost no one in Europe – except for Mediterranean sailors – used maps in 1400, maps were essential to a wide variety of professions, not least that of state rulers and bureaucrats by 1600 (Buisseret 1992: 1). This was also the case in Denmark. Analysing the process of mapping Danish territory according to the principles of scientific cartography in this period, I will emphasize the agency of map-making and how the novel mode of mapping established a new spatial reality altering the conditions for territorial governance. During the period 1448–1660, the Danish realm underwent a transition from a fairly decentralized *ständestaat*, constituting a union with the other two Nordic countries, Sweden and Norway, and living a fairly isolated existence in the Baltic region on the fringes of the Holy Roman Empire, to a centralized state implementing a universal set of laws and becoming an integrated state in the European states system. Indeed, in 1660 Denmark turned towards absolutism and many of the centralizing reforms associated with the modern state were instigated in the following 50 years.

The attempt by the Danish state to gain authorship over the cartographic representation of its territory can be considered completed with a novel, and fairly accurate map, ten years prior to absolutism. Hence, the cartographic unification of the territory was in place prior to the formal centralization of state authority, and the subsequent reforms to centralize and deepen control by the state. Although not signifying a causal relationship between cartography and the political transition to a unified rule under a hereditary king, the history of Danish state formation and mapping supports the notion that the map precedes the territory. More importantly, though, this story signifies that it was the establishment of space as autonomous through a particular mode of cartography that allowed space to become the defining dimension of sovereignty. Without establishing space as seemingly autonomous from social practices, it would not have been possible to establish space as the key marker of inside/outside the political realm of the sovereign.

Denmark, as an absolutist state, denotes a sense of unified political agency. Even in guise of an absolutist state, however, Denmark was a state composed of the Duchies Schleswig and Holstein, Norway, the Atlantic possessions Greenland, Iceland and the Faroe Islands, and the area *Trankebar* in India and some trade stations on the 'Gold Coast' of West Africa. While these were all parts of the realm, there was, at the same time, a notion that these were indeed different parts, and in the analysis I will focus on Denmark proper and only make cursory references to the mapping of other parts of the realm. While this risks re-producing the overtly national perspective which the historiography of cartography in Denmark has suffered from, it avoids unnecessary complexity which would occur by telling a more general story of the entire region. While this would have its merits in an analysis of the relationship between different parts of the kingdom, between centre and periphery, it would not add to the overall argument which is presented as follows. To begin, I present Denmark and the notion of territory that was in place prior to 1450. Then, I investigate pre-geometric mappings of territory and emphasize how the cartographic transition involved a transition in spatial knowledge from a literary mode to one in which space became autonomous and gained its own standardized measure. Finally, I analyse the mapping process of the Danish territory leading up to 1650, and emphasize how the state began to interfere and gradually achieve control of the mapping of the realm leading towards centralized and spatialized understanding of the sovereign state.

Pre-cartographic territory

Not more than a hundred years after nearly being dissolved – the country had no king in the period 1332–40 and most of the territory was mortgaged to German princes and mercenaries (Bagge 1999) – Denmark emerged as the dominant power in Scandinavia during the fifteenth century. In 1448 Christoffer af Bayern died. He had been elected king by the three independent councils of Norway, Sweden and Denmark in 1442, 1441 and 1440 respectively. The year of his death signalled change on many fronts. Running against the ideal of the union, the Danish and the Swedish councils elected different kings and the main priority for the Danish king – Christian I – in subsequent years was to re-establish the union. As such, international politics in the Baltic was dominated by a rivalry between Denmark and Sweden. However, this was not only a rivalry between two territorial states fighting for supremacy within the 'union-area'. It was as much a conflict between rule by the aristocracy or by the king, between union and state. This struggle continued until the 1530s when Denmark-Norway and Sweden-Finland emerged as composite territorial *ständestaaten*. The two states finally recognized each other with the Treaty of Brømsebro in 1541 (Gustafsson 2000: 322; see also Rasmussen 2007).

These events took place in a European context where the Ottomans had conquered Christian Constantinople in 1453 and the two universal powers – Pope and Emperor – were in decline. Germany, under the Empire, moved towards a federation of small states ruled by the Habsburgs from 1438 (Due-Nielsen et al. 2001: 167). Generally, one could argue that this was a time of spatialization, or territorialization, of politics; not only in terms of the emerging state system, but also with an increasing spatial division of the world. As shown in the previous chapter, the Treaty of Tordesillas of 1494 divided the world into a Spanish and a Portuguese half. Simultaneously the globe was divided into a new and an old world signalling the reconfiguration of established knowledge categories. Christianity came to be equated with a Europe defined in spatial terms thus demarcating the universality of Christendom (Mignolo 1995). And the significance of non-territorial polities decreased vis-à-vis the territorial political entities.

In this process, Denmark was located at the margins of Christian Europe and it was only slowly integrated into European politics. The 'international relations' of Denmark in the period changed from being part of a fairly self-contained Baltic Sea region on the fringes of the Holy Roman Empire, to being an integrated part of the European state system. Denmark contended with other territorial powers, but for years its main adversaries were the Hanseatic League of city states and the Teutonic order, both representing different territorialities, as well as, autonomous areas such as Dithmarschen in Holstein which was only conquered in 1559 after a spectacular failed attack in 1550 (Netterstrøm 2007). In this process, the role and meaning of territory changed, as did the way it was controlled and the way in which it serves the state as a means of power and its source of identity.

The lack of 'scientific mapping' in medieval Europe did not mean that there was an inadequate knowledge of space, or that there was no conception of territoriality, but rather, that knowledge of space provided different possibilities for the mode of territory. As argued by Monica Smith, the cartographic depiction and analysis of both ancient and modern political entities stems from a modern view of the necessary interdependence of political authority and physical territory. Generally she proposes that states should be analysed 'as networks of resource acquisition in which territories and their boundaries are porous, permeable, flexible, and selectively defended' (Smith 2005: 835). In support of this she notes how the Incas prior to the arrival of the Spanish, 'understood or conceived of their domain through roads, and not through provinces' (Smith 2005: 840). This is an important point but even where the relationship between political authority and territory was not defining, there were still a conception of territory emerging in medieval Europe. In the case of France, Jacques Revel has argued that around 1300 'the king had a legal perception of his territory and above all a fiscal one' (1991: 134). Generally, territory represented a source of income, both directly as farmland and taxation rights, but also because control of territory permutated the constitution of things such as trade rights. In Denmark there was an established perception of territory in the thirteenth century. The clearest evidence of this is King Valdemar's Cadastre (*Kong Valdemars Jordebog*) from 1231 which provided the earliest known surveyance of the land in Denmark and was undertaken for taxation purposes (Ulsig and Sørensen 1981). It contains a very specific description of the tax obligations of each shire and the market towns (*købstæder*) and, furthermore, lists the size of cultivated land in each shire as well as a description of the king's demesne (Fenger 2000).

Christianization (around 965) brings Denmark in touch with institutions that secure some permanence to, at least *de jure*, authority. At the same time, the king entered a changing relationship with the emperor sometimes as a vassal sometimes claiming autonomy. Though, from the late twelfth century the king denies vassalage to the emperor and establishes himself as an autonomous ruler (Due-Nielsen et al. 2001: 67). Subsequent territorial expansion in North Germany led the king to adopt the title *Nordalbingie Dominus* – an indication of dominion which is confirmed by a papal bull in 1214 – after which such territorial acquisitions *de jure* became part of the Danish territory. Denmark thus entered a world in which territorial rights were granted by the Pope and Emperor. The description of these rights was 'literary' and the notion of the territory was preserved within the textual knowledge tradition of Catholic Europe. There was thus a clear written account of the territory, its divisions and obligations to the king during the Middle Ages. This shows that it was possible to have a notion of territory and its extension without a visual representation of it.

What should be emphasized is that the knowledge of the territory was produced as a written text in both the cartographic traditions most widespread at the time: the *mappaemundi* and the *portolans*. Neither of them were initially concerned with territorial demarcation; the former tradition generally aimed to organize space according to religious and philosophical principles. The medieval world maps aimed to portray universal history as well as historical space (Edson 1997: viii & 15). Thus, the *mappaemundi* provides the cosmos, the universal Christian order, that Denmark became a part of. While the *portolans*, on the other hand, were initially illustrating sailing routes they gradually merged with the *mappaemundi* traditon and formed the basis of world maps. Several of these started to show flags as symbols of royal authority on the maps. This style of *portolans* thus came to visualize a link between the territory and royal authority and illustrate an internationally recognized political order during the fourteenth century. They were not made for any practical purpose, but were made to illustrate the order of the world and cosmos. And in this context, the Danish territory had come to be identified with the symbol of the Danish king. However, where these maps illustrated 'an international order' they did not provide much of a tool for the state to control the territory. Practical spatial matters, such as property dealings and tax collection, were still based on textual description and mediated through personal networks.

As space was not made present through its visualization, it could only be described with reference to already shared knowledge of spatial locations. Without a map, space and directions in space can only be communicated if another shared spatial reference exists and the literary descriptions required an extensive use of words. In an example from medieval Marseilles, Daniel Lord Smail describes how

> every one of the hundreds of thousands of property conveyances extant from medieval Europe included a clause identifying the location of the property by means of words alone. When, in 1352, a citizen of Marseilles named Ayvart Valhon purchased a house on the corner of two public streets, the notary who transcribed the transaction conscientiously named the two streets – 'the avenue of the [gate of the] Frache and a street or set of steps which leads to the church of St. Martin'.
>
> (2000: 6–7)

These addresses were mapped linguistically rather than graphically. More generally, in the context of Marseilles, spatial knowledge was mediated by notaries who were dealing with legal contracts and property relations. Yet, there existed no universal cartographic template and the development of the 'notarial culture' was not caused by the state. According to Smail, 'the public notariate, continuously developing in Europe from the middle or late twelfth century onward, was not an organ of a particular state. The legal genealogy was rooted in a *ius commune*, and notarial culture itself was heavily shaped by the usage to which it was put by consumers of the law' (Smail 2000: 226). In effect, spatial knowledge was dependent on personal relations and could not be abstracted from its social milieu.

In a similar fashion, during the sixteenth century, the Danish state was dependent on parish priests for obtaining local spatial knowledge necessary for tax and administrative purposes. There was no universal template for the terminology or classification principles of the peasant population (Ladewig Petersen 1980: 24) and, in effect, the state relied on a network which provided unstandardized knowledge which, in itself, was dependent on the persons who provided it. With regard to the measure of land, the categories were derived from practical and daily experience and, characteristically, these were not uniform measures in terms of area, but also reflected quality of soil, shape and similar qualities significant for those who lived off the land. The lack of standard measures and knowledge templates made it very difficult for the state to control and compare the information and thus to establish a uniform administration of the territory. Such standardized knowledge was obtained in relation to tax reforms and a new land register following the introduction of absolutism in the seventeenth century (Ladewig Petersen 1980: 24).

As a consequence, the representation of space did not provide a knowledge that allowed for sharp boundaries to be drawn or that enabled the uniform organization of large territories. Medieval political territoriality is usually described as being overlapping and hierarchical with no clearly defined centre and with ambiguous boundaries. Territory did not play the same role in defining the domain, but followed as a result of what Sahlins termed 'jurisdictional sovereignty' which was predominantly a relation between the king and the subject (1989: 28); and in the words of Rhys Jones, '[t]he key significance of territory, therefore, during the early modern period was its use as a spatial manifestation of the notion of sovereignty' (2007: 21). This means that it was the relationship of loyalty between, for example, feudal lords and the king, which decided the extent of the kingdom. And as ownership of the land from the 1200s onwards became increasingly concentrated around the nobility and ecclesial elite (Hybel 2003: 276–7), it was largely the loyalty of these elites that would have to be secured in order to maintain the territory.

As a *ständestaat*, Denmark was organized according to a semi-feudal structure which, as Teschke states, is characterized by personal decentralized authority (2003: 214). To become king around 1500 in Scandinavia three things were required: one had to be elected by the council,

accepted by the estates and towns and have control over the castles.
The castles were the keys to controlling the territory and served as
centres of mediation extracting wealth from the peasant population and transferring it into royal possession (Gustafsson 2000: 45).
The castles were controlled partly through being owned by the king,
co-operation with the church, and with the castle-owning nobility who
had sworn allegiance to the king. These relations were generally personal and had to be renewed with the heirs when the nobleman died
(Due-Nielsen et al. 2001: 67). The territory was controlled through a
multitude of 'centres' and this meant, for example, that peace negotiations were about deciding which castles, fortifications and cities should
fall to one part or the other rather than drawing a border on a map.
Control and knowledge over the territory was, then, achieved through
a web of personal intermediaries.

As there were many 'centres' there was also no legal homogeneity. As
noted by Østergård, 'Denmark comprised a number of legal communities that ranked next to each other and only later were united in larger
constitutional units' (Østergård 2002: 10). Even though a notion of territory was established there was not just one, but at least two, competing
territorialities. The church owned a large number of significant castles,
owned about a third of the land and did not necessarily respond to the
king's wishes and, as administrative knowledge was mostly obtained
from the Parish priests, the administration was dependent on the
Catholic Church. Hence, Danish territory at this time was discontinuous.
Control was mediated and dependent on personal relations and controlled through various centres rather than one. Generally, then, space was
not a resource that could be manipulated by the state especially because
it was complicated to extract financial resources from the countryside.
During the period under discussion, fortifications remained key to defen-
sive strategies, but their role and their organization changed alongside a
general change in the knowledge of space.

Transforming knowledge of the territory

Little is known about medieval land surveys but there were no
standardized framework for organizing spatial knowledge (Price 1955).
Practical usage and units of time were often used for measurement.
Kong Valdemars Jordebog, for example, used the term *plove* (ploughs) for
tax purposes (Aakjær 1980). For the measure of land, there existed a
number of different forms but throughout much of Europe two forms
of measurement were common: a carrucate (*plovland*) denoted the

area which could be ploughed in one day and a barrel (*tønde*) of land indicated the area which could be sown with the seeds of one barrel (Kula 1986: 29–42). Although designated 'measures' of space existed,[2] these were not used for greater distances or areas at the time and there was no one single standard for measuring all 'spatial relations'. Instead spatial practice had to rely on different sources of spatial knowledge.

For example, there exists no historical indication that the Vikings produced any kind of maps (Edson 2005); instead they navigated with the help of a reckoning dial measuring the shadow (and thus the height) of the sun and other instruments which could help them stay on the same latitude (Kragh 2005: 35–6). In addition, sailing routes would be passed on orally and in written form. As indicated by the great distances travelled by the Vikings, this mode of knowing space worked well, but it was very much based on personal relations and it had no general template in which to assemble the knowledge. Longer distances were usually measured in travel time. Again, an example from *Kong Valdemars Jordebog* describes the route from Denmark along the eastern coast of Sweden via Aalandshavet to Finland and from there to Estonia. It contains a detailed description of the directions and distances between landmarks are given in 'weeks-at-sea' (*uge søs*) (Nørlund 1943: 21). In an example from France in the first half of the fifteenth century, the king's herald stated that 'the length of this kingdom is a twenty-two-day journey, from l'Escluse in Flanders to Saint Jehan de Pié de Port at the border of the kingdom of Navarre, and the width is a sixteen-day journey' (Revel 1991: 148). As geometric prescriptions of measuring space became a single standard they replaced a variety of 'functional measures' of space.

In producing knowledge of space according to the new rules of fifteenth century cartography, theory came to precede experience. That is to say, that the rules of space are defined prior to – and disembedded from – practical experience. Navigation of the fifteenth and sixteenth centuries saw a transition from experience-based mapping of the *portolans* to a geometric one where the rules and the framework of the map were given before any spatial data entered the map (Parry 2000: 100–113). Space thus becomes an 'a priori' category defined autonomously from the immediate experience of the environment and measured by the means of units abstracted from practical use. As discussed in Chapter 4, locations in space became a matter of the relationship between celestial features, the earth and geometry, and knowledge of the territory was transformed from a literary mode to a mode of visualization and calculation – supposedly – independent of practical experience. Establishing space as an 'autonomous' category, which in principle could be perceived without reference to functional time, the immediate experience of the environment, or the author of the map, enabled a uniform visualization of territory. This rendered it coherent and tangible in its own right and space became a 'thing' which could be possessed; not as a field, or a forest, but as a parcel of space. It became possible to draw boundaries in this abstract space only to effectuate them on the ground at some later stage – if at all.

Initially, the transition to scientific cartography was not led by the state, but rather by scholars and commercial publishers. At least partly, the transition seems to have been underpinned by an ethos of truth informed by a novel notion of how space could be truthfully represented. The enterprise was dominated by a private map-producing circle of publishers in Europe. Much of the demand came from wealthy individuals and they increased the demand for maps produced also in the international university environment. By the end of the fifteenth century, printed maps had become sufficiently popular to provide the basis for specialized map sellers and during the sixteenth century a more elaborate trade organization developed around Venice and later Amsterdam became the undisputed centre for map-making in Europe. The map-makers' workshops could be interpreted as centres of accumulation where spatial data was gathered and gradually added to the new global template of spatial knowledge. Danish map-makers entered these networks of spatial knowledge production and provided local knowledge to the centres.

Although a few Danish place names were included in the standard Ptolemy text, it generally contained little knowledge about the Nordic countries. Such knowledge was added, however, by the already mentioned Claudius Claussøn Swart, who produced a map of Scandinavia which quickly became part of the standard Ptolemy edition. Swart explicitly writes himself into a tradition where he, as an eyewitness, possesses knowledge unknown by 'the authorities'. In one of the key texts, he states that 'I, the Dane Claudius Claussøn Swart, [...] have by meticulous drawing as well as written records aimed to give to the world a true picture of said countries known to me by personal experience, yet unknown to Ptolemy, Hipparch and Marinus' (quoted by Nørlund 1943: 13, my translation). The significance of Swart being from Denmark is not that he added to a Danish cartography, but that this gave him first hand experience which he could add to a general European body of knowledge with universal scientific aspirations.

The other significant map showing Denmark in the early period is Olaus Magnus' *Carta Marina* published in 1536 (see map 6.2). It became very influential for the production of subsequent globes and world maps in Europe. Magnus was of Swedish origin and, like Swart, went to Rome

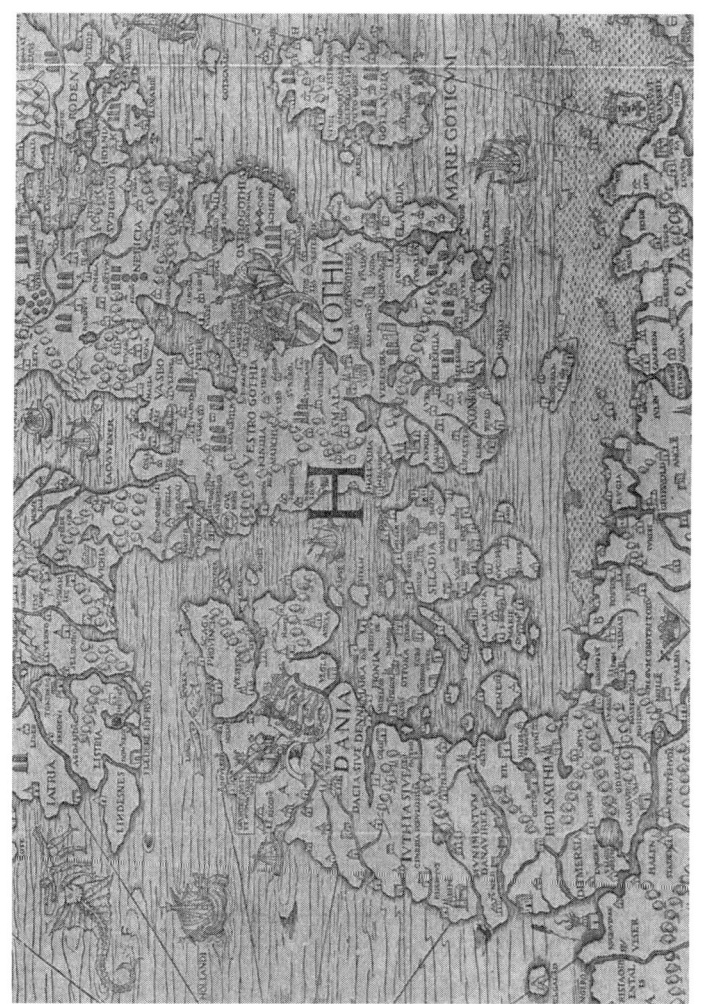

Map 6.2. Section of Olaus Magnus' *Carta Marina*. Notice the two kingship positions signalling authority over the land. This is a 1539 edition. By courtesy of the Royal Library Copenhagen.

where he spent much of his life. Both Swart and Magnus were part of an international network of scholars who communicated and copied each other's work without working for any state in particular.

Apart from the two examples of Swart and Magnus, which were both published in an 'international context', the cartographic representations of the realm were made by 'outsiders'. Or rather, territory was mapped and published by the map-producing networks discussed in the previous chapter. Swart contributed explicitly to the Ptolemy editions and both he and Magnus were influential for subsequent depictions of the Nordic realm in the European map industry. In the previous chapter we discussed how the Netherlands, during the sixteenth century, became the centre of map-making at a time where it also acquired a leading role as the economic centre of Northern Europe. As such, the Dutch cartographers in the sixteenth century generally possessed the power of authorship deciding the cartographic image of the Danish territory. The cartographers' workshops played the role of hubs, or centres of calculation, in the networks of spatial knowledge production concerned with a universal cosmographic enterprise. It is due to this network that we know of the first map to be published in Denmark. It was made by Marcus Jordan in 1552, where he was professor in mathematics at the University of Copenhagen. This map was not based on his own surveys, but was to a large degree based on previous sea charts and the influential *Carte van Oostland* published around 1550 (Nørlund 1943: 26 & 42). Taking its place in the European cartographic network, Jordan's map provided the basis for the map of Denmark in Abraham Ortelius' *Theatrum Orbis Terrarum* (see map 6.3).

The personal networks underlying cartographic practice are also clearly noticeable in the other great atlas: Mercator's *Atlas sive Cosmographicæ*. Heinrich Rantzau, who occupied a leading position within the Danish nobility as head of the king's administration in the Duchies was also interested in the new geographic disciplines. He was behind the first known 'Danish map' of Denmark, as discussed below. Establishing contact with Rantzau, Mercator promised to give Denmark a prominent place in his atlas and that 'the glory of your name will be celebrated from help of this kind' (quoted by Crane 2002: 267). And Denmark is indeed described in very detailed and glamorous terms. In an English translation from 1636 we read of the 'great and populous Kingdome of *Dania*' that '[s]uch was the forme of the ancient Politicke State , & the Monarchie of the Danes; which never any nation was able to conquere, nor make them loose their ancient priveleges, and customes' (punctuation kept as original, Skelton et al. 1968: 95). The text contains a detailed description of the political

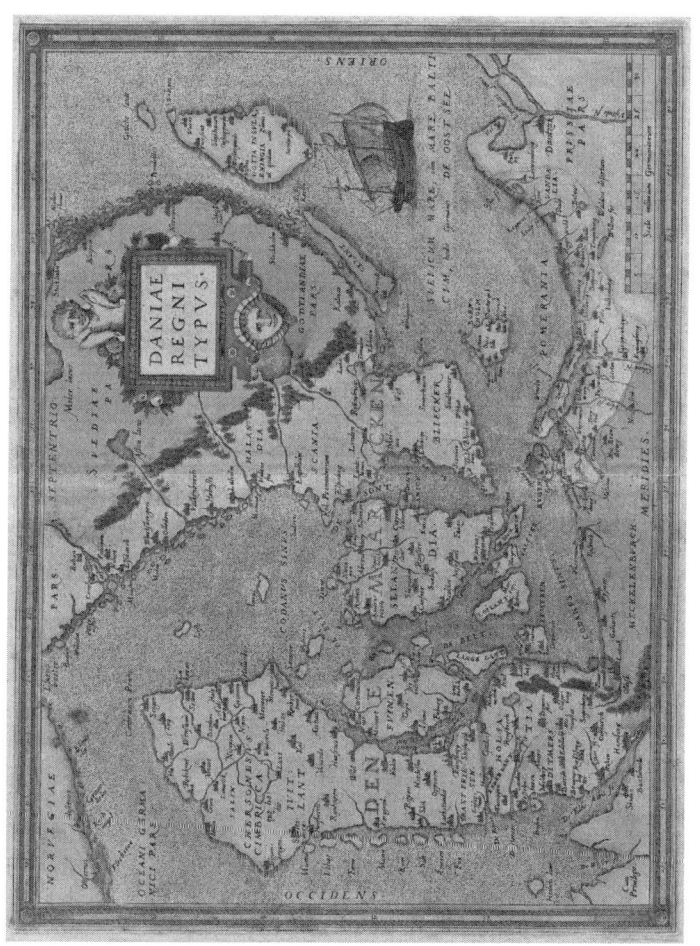

Map 6.3 Ortelius' map of Denmark from *Theatrum Orbis Terrarum*. This is a 1572 edition. By courtesy of the Royal Library Copenhagen.

structure and customs, as well as numerous references to the beauty and richness of the country. It also presents several accounts of the deeds of the Rantzau family (Skelton et al. 1968: 101). Due to the connection between Rantzau and Mercator, the atlas contains 4 different maps of Denmark, whereas Sweden is represented with a single map. Not unexpectedly, the 'extra maps' show the areas where Rantzau had most of his possessions. In this respect, Denmark now occupied a significant position in the publication of the emerging globalized and unified space of the world. Various scientists and noblemen, like Rantzau, were involved in putting the state of Denmark on the map. However, Danish territory was represented as part of a universal enterprise to map the world, and thus not in a fashion controllable by the monarch. This, in itself, might have provided an incentive for the sovereign to seek control with the authorship of his own territory. But no less significantly, a more controlled and detailed mapping of the territory would provide new possibilities for the state to develop its territory.

Assembling Danish territory

In Denmark, the situation was unstable after attempts by Christian II to enforce the Nordic Union under his kingship. During his crowning ceremony in Stockholm in 1520, he had set a trap for a large part of the Swedish nobility that were executed; possibly in order to quell their resistance to the Union. The result was adverse to Christian's efforts. Swedish resistance grew increasingly organized and, in 1523, he faced rebellion 'at home' as well, and he was toppled as Danish king. He had been trying to expand his power base in the peasant and merchant estates and thus tried to limit the influence of the nobility. Christian left Denmark and took the entire fleet with him as his personal possession. Christian's uncle, Frederik, was elected to succeed his nephew as king, but ruled under a constant threat of an attack by Christian who, as son-in-law of the Emperor, still had a legitimate claim to the throne and very good connections in Europe (Due-Nielsen et al. 2001: 220–60).

Following an unsuccessful attack on Norway, Christian was snared into captivity. This did not, however, resolve the tension as after the death of Frederik I the council postponed the election of a new king. The interpretation of the following events are widely disputed, though, in 1534 the city of Lübeck attacked Holstein to get Christian II released from captivity and reinstated him as king. This attempt was largely supported by the town and the peasant population whom Christian II had relied upon against the nobility. However, eventually the nobility and their candidate for king, Christian III, emerged victorious obtaining control of the entire country in 1536 (Due-Nielsen 2001: 272). As a result of the civil war and the events surrounding it, the Nordic Union was left as a dead project in the face of two territorial states Denmark(-Norway) and Sweden(-Finland).

Christian III was a 'modern style' European sovereign who had lived and been educated in the German territories, and significant changes followed after his accession to the throne. The new regime pursued a centralization of power, the Catholic Church was dismantled and the Catholic bishops, who until then had been the most powerful fraction of the council, were charged with high treason. The king became head of the new Protestant church and this brought an end to the alternative territoriality of the church that had existed in parallel with royal authority since Christianization. Furthermore, large-scale fortification projects were initiated throughout the country and the administrative structure of the fiefs were reformed to pursue standardization and centralization. With the introduction of accounting it was possible for the central administration to regulate and audit the local administrations (Jespersen 2007).

The vassals were gradually transformed from being an independent nobility, with certain duties to the king, to the king's representative in the local area. Along with the new notion of sovereignty, the expansion of the central administration fostered a notion of shared interest between the nobility and the king (Jespersen 2007: 73–8). In connection with this, a law was passed (*Slotslovene*) which gave the king command over all castles in the realm (Due-Nielsen et al. 2001: 273–7). New fortifications generally pushed the defences towards the boundaries and the previous castle-owning nobility moved into residences of splendour rather than defence. The king thus monopolized the defence of the realm under the auspices of the crown and, generally, these trends followed the European fashion where, in the words of John H. Elliott, the word state 'used to describe the whole body politic, seems to have acquired a certain currency only in closing years of the [sixteenth] century' (2000: 49). These developments helped to institutionalize a notion of territory, which was becoming increasingly independent of the personal network relations between the king and the vassals. As the nobility became royal bureaucrats, and the castles lost their administrative function, the territory was increasingly governed from the capital. This was also reflected in fact that the king travelled less and less during this period. As it became possible to govern the territory from the centre, the king's presence was less required in the different parts of the kingdom. Hence, a shift occurred from what could be labelled a 'travelling' to a 'centralized' kingship (Jespersen 2007: 73).

With the restructuring, sovereignty was no longer 'unilaterally bound to one or other of the central institutions of power – the monarchy or the state council – and was instead linked to an impersonal, abstract and permanent concept, that of 'The Crown' (Jespersen 2004: 34). This represented a depersonalization of the notion of sovereignty, and the abstraction and continuity embraced by the notion of the Crown facilitated 'the state' to exist by virtue of itself. It was this notion of sovereignty gradually abstracting from the king's body which would be attached to a specific territory (Neocleous 2003: 410; Bartelson 1995: 98), and hereby promote the spatial identity as a state.

The adoption of cartography in government slowly started in the fifteenth century but broader developments of this practice were slow. For the most part, it was only during the sixteenth century that maps started to play a significant role for the European states (Buisseret 1992: 2), and this was also the century where several 'national' mapping projects were ordered by sovereigns. Following this trend, in 1553 King Christian III asked Marcus Jordan to map 'all the kingdom's provinces, islands, towns, castles, monasteries, estates, coastlines, capes and anything else worth noticing' (from Bramsen 1975: 52, my translation). The result of Jordan's engagement by the King is not fully known. He only produced maps of some of the regions in Denmark and the map he delivered to the King has not survived. But, in 1588, a map by Jordan appears in Braun and Hogenberg's *Civitates Orbis Terrarum* (see map 6.4). It was printed in the fourth volume of this first city atlas mirroring the title of Ortelius' atlas. The contribution by Jordan had been ordered by the governor of the Duchies, Heinrich Rantzau (Ehrensvärd 2006: 100), and the result was that Jordan produced the first known map representing Denmark produced in Denmark. It was based on Jordan's own surveying – the job he was employed to do ca. 30 years earlier by Christian III. Yet, it was not the King but a powerful member of the aristocracy who provided the means for the map to be produced. This is possibly the reason that it does not represent an 'even picture' of the realm but rather a tribute to Heinrich's father Johan, who famously crushed a peasant rebellion during the civil war that brought Christian III to power (Venge 1981), and is a testament to the extension of his lands. Where the map is most detailed (Fünen, the Duchies, and northern Jutland) is also where Rantzau owned plenty of land.

Compared to previous records of the territory, this map provides a novel representation of the realm, showing boundaries as neat lines

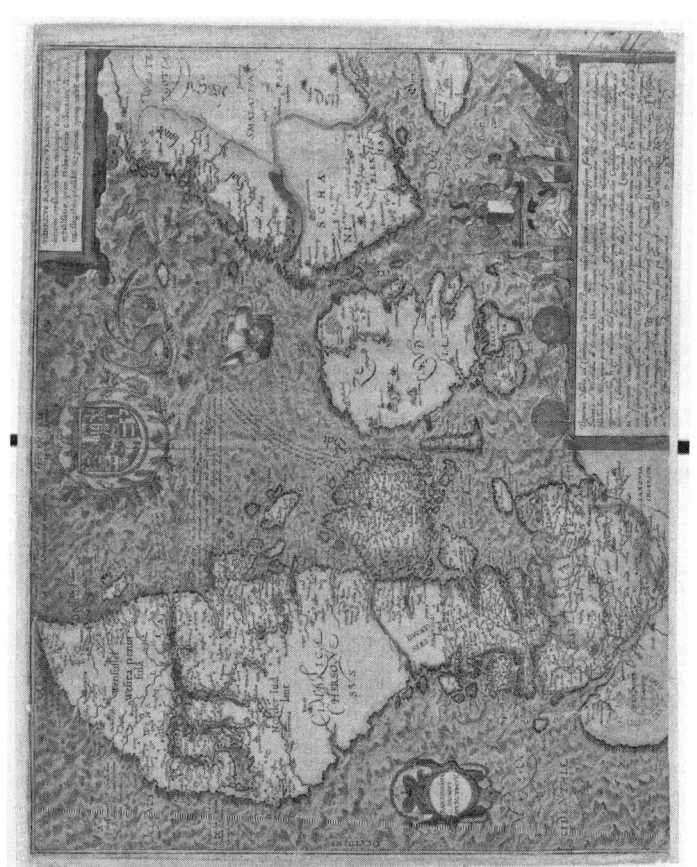

Map 6.4 Jordan's 1588 map of Denmark from Braun and Hogenberg's Civitates Orbis Terrarum. By courtesy of the Royal Library Copenhagen.

and containing a brief description of the composition of the country. The central display of royal arms, as argued by Helgerson, speaks of a relation of power to the land (1992: 111); it signifies the ownership of the King over the entire territory and renders, in theory at least, the nobility and the estates invisible. Hence, this map both re-affirmed the territorial aspirations of the King, and the grandeur of the Rantzau family.

As the geometric map reduces the complexity of the world and abstracts space from its 'social reality', it provides an alternative reality; in this case one where the territory could be monitored in its totality on a single sheet of paper, and all locations, coastlines and so on can be viewed and related to each other in an instant – also by strangers who had never visited the country. In *Kong Valdemars Jordebog*, the description of the territory was also available to all literate people, but the listing of islands, for example, would not give much of an idea about their location, shape and size unless you knew them already. On the map, the location, the name and the shape were all linked and instantaneously present. Nonetheless, Jordan's map was still not a great success in terms of creating a uniform and detailed map of the kingdom. The map did not appear until long after the King's death (1559) and it was published in Hamburg, which had an ambiguous relationship to the Danish state. More importantly, it was published as part of a general description of 'the cities of the world', contributing to a general body of knowledge. Hence, although a new representation was being established, a unity between authorship and the crown was still not achieved.

Around 1580, a new project was launched that, involving Tycho Brahe, was supposed to provide a new and improved map of Denmark as part of the history of the country (Kragh 2005: 285). This was supported by the King Frederik II, who ordered that Brahe should have access to all the maps held in the King's library. Surrounded by a group of people that became prominent in the mapping of all the Scandinavian and North Atlantic countries (Mead 2007: 1790), Brahe's activity illustrates yet again the network structure of European cartography at the time. There were connections and exchange of knowledge between Brahe on Hven, Ortelius in the Netherlands and Gudbrandur Thorláksson, who significantly influenced the Mapping of Iceland and Greenland at the time (Kejlbo 1980: 200). Brahe was the first to utilise the method of triangulation, which later became standard for scientific map-making. The only direct result of this, however, was a very precise map of the island Hven, which was in the possession of Brahe and where he had build his workshop and castle Uraniborg (Nørlund 1943: 45–6).

After the accession of Christian IV in 1588 to the throne, tensions grew between Brahe and the king and the disputes eventually made Brahe leave his island fief. While the king was reluctant to fund Brahe's research unconditionally, the state generally became more active regarding cartographic activity in the attempt to produce an accurate cartographic representation of its territory; not only in Denmark but also the rest of the realm where Christian IV began sending expeditions to Greenland to 'rediscover' the island, assert Danish sovereignty, and find new sources of wealth (Keller et al. 1967: 57–62; Kejlbo 1980).

In 1622, the state employed Dutch copper engravers for map publishing and in 1623 another professor of mathematics, Hans Lauremberg, who was teaching geometry, surveying and the art of fortification at *Sorø Akademi*, was appointed to map the realm. Lauremberg was granted access to all areas of the kingdom, and received a regular payment from the state. The king followed the process closely and, in 1639, the head of the financial administration, Corfitz Ulfeldt, received an order to make sure that Lauremberg's maps were engraved and published (Nørlund 1943: 48–50). Despite this effort, the project was never completed and the maps never printed – at least not in Denmark. A likely reason for this was Christian's insistence on publishing them in Denmark under his control, whereas Lauremberg wanted the Dutch master publishers to do the job.[3] Lauremberg lost the King's favour, and the King's attempt to fully control the process had failed again. Just before Christian IV's death, the task was handed over to Johannes Mejer and his small crew, who were given a deadline of 6 years in 1647. This attempt was more successful. After his appointment he produced a large number of maps and handed over a general map of Denmark to Frederik III, who had succeeded his father as king (see map 6.5). The map was printed in 1650 and, although never generally published (Nørlund 1943: 53), finally provided the king with a uniform map of the territory and he had thus become 'the author of his own world'.

As mentioned earlier, this was not a map of the entire territory of the Danish state. In that case it should have included Norway, Iceland, Greenland, the Faroese Islands, various islands in the Baltic Sea, such as *Gotland*, as well as the small trading colony *Trankebar* in India. But it covers what was considered 'Denmark' within the realm. Similar mapping projects took place in the other parts of the state as well, sometimes by other cartographers, and sometimes by the same. Johannes Mejer, for example, had an ambition to complete a Scandinavian atlas and undertook surveys in Norway and published new maps of Iceland and the Faroese Islands (Nørlund 1944a: 33, 1944b: 18; Mead 2007).

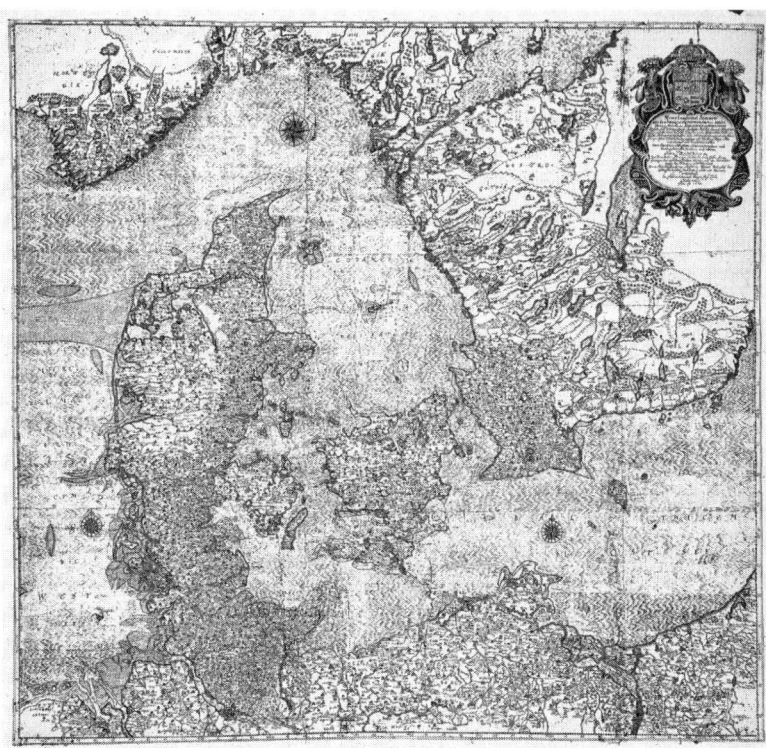

Map 6.5   Johannes Mejer's 1650 map which was very accurate in its geometric representation. By courtesy of the Royal Library Copenhagen.

With these caveats in mind, Mejer's map, even if it did not live up to subsequent scientific standards, provided the state with a novel guide to the territory. As such, it completed the transition from a literary mode of knowing the territory to a uniform cartographic one which could abstract from local knowledge and combine – assemble – these into a coherent framework at the centre. This contributed to a centralization of knowledge of the territory, and hence, unifying authorship and the state (Turnbull 2000: 116–17). The process of mapping the Danish territory has illustrated that the production of a science-cartographic representation of territory was no straightforward process or one strictly controlled by the monarchs. Hence, one should be careful about naming a specific moment from which the new cartographic representation of the territory was complete. Yet, it is noteworthy that Mejer's map was printed ten years prior to the introduction of absolutism because this suggests that the cartographic unification of space preceded the centralization of sovereign power.

Although not initially controlled by the state, the 'cartographic revolution' made it possible for the state to become an abstract spatial body, independent of personal relations. Now that the entire territory was available to the observer in an instant it was no longer necessary to travel for weeks to know the territory. Borrowing the words of the famous French cartographer, Nicolas de Nicolay, who was commissioned to map France by Catherine de Medici: '[i]n little time, in little space, and without great expense, to see with the eye and trace with the finger, in particular and in general, the whole extent, power, and status of the realm' (quoted and translated by Serchuk 2006: 146). Or in the dedication from the Swedish cartographer Andreas Bureus we read: 'This little map I thee give instead of the whole world, and then I shall give thee the whole of Scandinavia on a copperplate' (Mead 2007: 1794). The unification of authorship and the state completed the transformation from a literary mode of knowing the territory to a uniform cartographic one in which territorial space could plausibly be claimed by the sovereign as an abstract space.

Centralizing authorship and absolutism

As space was established as autonomous it became possible to identify the state primarily in territorial terms. It is the map that provides the state with its spatial identity because it is through cartography that state and space are merged into one representation claiming to be the truth about space. The state could now 'become a self-referential' political body as it could point to the map and say 'this is me/mine' (Harste forthcoming). Or, as argued by Helgerson, early mapping of the state in Europe made possible effective visual and conceptual possession of 'the physical kingdom' (1992: 107), and maps thus provided the state with a spatial identity. It also provided a novel spatial setting for historiography. For example, the theology professor, Erasmus Lætus, published *Res Danica* around the same time as Jordan was carrying out his surveys (Wolff 1979: 379). *Res Danica* was not only a history, but also contained a description of localities and buildings around the country and is thus an example of linking history and geography supporting the notion of a state with a continuous and stable spatial identity. During the subsequent decade, Frederik II ordered 43 large tapestries depicting more than 100 Danish kings purposively indicating the ancient historical roots of Denmark. The last tapestry showed Frederik II with his son Christian thus representing an unbroken line of kings from a mythical past to the present (Heiberg 2006). The tapestries were ordered with the purpose of decorating the Knight's Hall in the newly built Kronborg castle, protecting the passage through the Sound to the Baltic Sea and ensuring the largest source of income for the Danish state: the Sound toll. This is to say that the geometric map provided new conditions for how the state and its history could be 'written into' a settled bounded territory, and this merger between the state, history, and space legitimized sovereign claims to authority over a specific and neatly demarcated piece of territory.

Especially following the end of the civil war in 1536, centralizing reforms were set in motion under the rule of Christian III, but with the introduction of absolutism, this process was significantly accelerated and many of the institutional structures which characterize the modern centralized state were inaugurated during the reforms that followed in the years immediately after. Many of these reforms, I argue, depended on the new mode of establishing the reality of space. Hence, if the rise of absolutism can be seen as a precursor to modern territorial rule,[4] then the novel mode of knowing space conditioned and preceded modern territory. After the defeat to the Swedish armies in the wake of the Westphalian peace treaties, the old political order was abolished and absolutism was introduced in Denmark in 1660 where kingship was made hereditary and the population, in principle, constituted as equal subjects (Bøggild-Andersen 1971). The king's 'new' authority descended from God – rather than the estates or the council – as a curious parallel to the new spatial order based on the rules of geometry and observation of celestial objects. This is illustrated well with the following quote from a bishop's speech at the crowning ceremony:

> Everybody must submit to those authorities above him as there is no authority not given by God and those that exist have been ordained by God ... Give to all what you owe them; to the one you owe tax: tax; to the one you owe toll: toll; to the one you owe fear: fear; to the one you owe honour, honour.
>
> (quoted by Ladewig Petersen 1980: 394–5, my translation)

In that respect both the political and the spatial order derived its measure from the sky and were in principle abstracted from the social networks that had previously served to constitute both space and royal authority.

The novel mode of knowing the space of the territory not only facilitated a novel conception of the state, it also equipped the state with practical means to enhance its capacity for resource extraction and centralized administration. In the words of Turnbull, '[m]odern cartography is the product of joint processes of cognitive and social ordering resulting in the establishment of the knowledge space within which scientific knowledge is assembled and the state is organized' (2000: 92). Following Latour, maps do two things: they store and organize knowledge, and they enable 'sending out' scientists from a centre of calculation and take disparate places back home and combine them with what already exists (1987: 219–21). Due to the way in which the scientific map stores and organizes knowledge, it became possible for the state to develop 'its space' into a territory proper. All information could now be fitted into one framework which made possible the assemblage of non-uniform pieces of information into a coherent image.

The networks in which the work of earlier cartographers such as Jordan and Ortelius evolved relied on a support network of royal and noble courts around Europe, tying the work of the cartographers back to the interest of powerful noblemen and sovereigns. The role of the courts, as noted by John Robert Christianson, was to provide the material means, and sometimes also the demand, for the natural philosophers, and during the time of Tycho Brahe and Marcus Jordan there were several important courts in Denmark serving as centres of accumulation in addition to that of the king (Christianson 2006: 1–6). The key in these relationships were the more or less formal networks between surveyors, cartographers, and the courts under whose auspices they were often working. The increasing involvement and domination of the state in these relations represented a centralization of knowledge of the territory, and a gradual decrease in the importance of noble courts vis-à-vis the royal court. It was achieved by deploying surveyors from the centre who would bring back the different places of the territory in the form of spatial knowledge. Resembling the cartographer's workshops discussed in the previous chapter, the capital came to work as a centre of calculation where the growing body of surveyors and cartographers worked to improve the new cartographic image of the territory gradually adding new spatial data into a coherent whole (Latour 1987: 219–21; Turnbull 2000: 116–17). And this, for example, allowed central spatial planning of infrastructure and the improvement of rural taxation. Hence, not only did the map condition the centralization of authority, once in motion, scientific cartography provided the state with a tool to command space.

In the years following absolutism, a number of 'territorial development' initiatives were launched by the state. Following a four-year campaign to regain territories lost to Sweden in the aftermath of the Thirty Years' War, the state engaged in a comprehensive new survey of the countryside during the years 1681–7 to improve its ability to raise taxes

(Heering 1932: 14). The result was Christian V's famous Land Register which provided a uniform registration of the entire territory (Henriksen 1971: 10). Traditionally, the Crown had three sources of income apart from the royal demesne: toll, fief payment and taxes. It was a very costly and arduous affair taxing the land compared to taxing easy controllable trade routes (Jespersen 2007: 73). Yet, with the new surveying and mapping techniques it became possible to develop an administrative structure improving the taxability of the land. This was done in France where, according to Turnbull, cartography was seen as the solution to the problem of finance (2000: 113). Indeed, a 'cartographic description was seen as perfect measure of the identity of the land. To some, it seemed that the certainty of property could be assumed only by delineating 'land' on a proper surface of description' (Alain Pottage quoted by Turnbull 2000: 103). Thus, by changing the record of private property to a systematic cartographic one, geometric cartography changed the possible tax base of the country. The traditional functional measure of land, barrel (*tønde*), was retained, but, with the new land register of 1683, it was standardized to cover 14,000 square *alen* (Equivalent to 5516 m²) (DSDE 1994: vol. 19). The measure of land changed from a functional definition to an abstract metric definition, thus facilitating the taxation of land.

Such standardized reforms were made possible by the state now 'possessing' the authorship of the cartographic representations, and the capital Copenhagen becoming a centre of calculation which allowed the state to expand control and 'use' space as a medium of power. In 1685, a commission was established to make a plan for how to improve and expand the connection between different parts of the country by 'king's roads' (*kongeveje*) reserved for the king and his staff (Jørgensen 2001: 26–7). These were the first centrally planned roads in the country. As a final example, a commission to create a new single legal code was established 1661. The law (*Danske Lov*), which unified the territory of Denmark proper excluding, the Duchies, Norway and the Atlantic possessions, was signed by the king in 1683 (Scocozza 2003: 315–19), thus unifying the territory through a universal code of law. All these examples of a significant movement towards modern political territory were preceded by a cartographic unified production of Danish territory. I would suggest that these movements would not have been possible without the spatial knowledge available. If this is true, it can be concluded that the formation of modern political territory depended on a very specific cartographic representation of space and, in sequence, that the development of the modern state was conditioned by the specific cartographic reality provided by the map.

Conclusion

In one of his lectures at Collège de France, Foucault presented a text from the seventeenth century by a Le Maitre who was employed as general engineer by the elector of Brandenburg. In the text, *La Métropolitée* he described the ideal geometrical order of a state. The foundation of the state should be the countryside populated only by peasants, the artisans should live in the towns, and the third estate, the sovereign and his officers, should reside in the capital. In addition, the perfect shape of the country was a circular territory with the capital in the middle (Foucault et al. 2007: 13–5). This story emphasizes the degree to which the geometric cartographic ordering of the territorial state had come to dominate political thinking. My aim has been to show how a transformation in cartographic practice changed the way in which territorial space was known, and how this altered the conditions for the identity of the state and the ability of the state to exploit the territory. New epistemic rules that came to govern European cartography from the fifteenth century onwards altered the reality of space and helped to establish it as an autonomous thing defined in terms of specific spatial metrics. With that it became possible to organize social life according to a predefined space; that is to say that it became possible to spatialize social organization. In consequence, it became possible to produce territory independently – in principle – of the social network which had constituted the territorial scope of the realm prior to the cartographic 'revolution'. Of course, social bonds were still a requirement to constitute the territory afterwards, but producing knowledge of the territory in the novel fashion implied that it became possible to abstract the coordination of territorial practices from their local settings, and coordinate these from the centre. In that respect the cartographic production of territory both conditioned the centralization, homogenization, and coordination of the territory as a political space – both on a representational and practical level. And in consequence, the transition towards geometric cartography provided the possibility of defining sovereignty in territorial terms.

The transition in cartography was not initially a result of state-coordinated practices. Prior to the sixteenth century, the geometric mapping of Danish territory took place within a European network of cartographers gradually assembling spatial data to complete a novel representation of the world. This process helped to demarcate the territory and fed into the imperative of exclusive territorial sovereignty. It served to define Denmark as a political actor on a global stage, but it did not in itself give the state authorship over the territory. Apart from the possible request by Erik af

Pommern to Claudius Claussøn Swart in the 1420s, it was not until the mid-sixteenth century that the state more actively pursued authorship of the territory. Several of these attempts failed in the sense that the new maps were still published in the Netherlands by cartographers who could not act independently of the Emperor's interests. This might, in itself, have provided an incentive for the ruler to claim authorship of the territory. And certainly, under the reign of Christian IV, who launched an early attack on the Emperor's armies during the Thirty Years' War in the 1620s (Due-Nielsen et al. 2001: 404–16), several cartographic initiatives were launched.

The map by Mejer, completed in 1650, signalled a completion of the state's quest to obtain authorship and within the following fifty years a number of territorial initiatives were launched by the state. Unifying the territory under a single centre of authority illustrates the possibilities provided by geometric cartography. Not least the uniform surveying and recording of the territory that was launched in the 1680s to improve the tax-base of the state, made all the more urgent by the attempts to regain lost territory from Sweden. However, the new land register did not produce topographical mapping due to a lack of resources and skilled personal. It was not until the Academy of Science produced their general maps from the 1750s onwards that we see a complete merger between scientific reason and the state. During this period, we are still a step away from the context where Harley concluded that, as 'cartography became more "objective" through the state's patronage, so it was also imprisoned by a different subjectivity, that inherent in its replications of the state's dominant ideology' (Harley 2001: 107). On the contrary, I focused on how the novel mode of establishing space as real conditioned a novel way of identifying the state and provided new opportunities for state leaders to centralize power. In this process they also came to dominate the authorship of space. This should be seen as a process in which space, established as a social phenomenon, conditioned state formation; and during this process the state came increasingly to condition the establishment of space.

7
Conclusion

Jeremy Bentham is usually accredited with coining the term 'international', in the context of law, in 1789 (Brown et al. 2002: 6). This was not only the decade of the French revolution but also when the first 'international space' was established cartographically. 1783 saw the first international cooperation, between France and England, when they determined the difference in latitude and longitude between the Paris and Greenwich observatories, enabling a joint mapping venture of the English Channel (Turnbull 2000: 121). Here, for the first time, surveyors from two countries came together in order to establish a common reality of space providing a stable and accurate scene for England and France. This signalled a transgression of space from being a national construct, as European states had achieved control over the authorship of maps, to an international one. The establishment of space was becoming an international effort where states contributed shared authorship. Eventually, Greenwich, reflecting British dominance worldwide, was agreed upon as a universal prime meridian in 1884 and hereby came to provide the basis of all measures of longitude as countries gradually adapted to the decision (Howse 1980: 116–72). Accordingly, the basic grid and measure of the earth left the realm of directly contested politics. Thus, agreeing on the standards of map-making, cartographic space came to represent the scientific spatial stage for international politics.

As described by Latour, the 'modern settlement' involved a firm distinction between what came to be considered 'nature' and 'culture' leaving politics on the cultural side of this divide. Remember, it was the 'insertion of a unified nature' behind cultural differences which constitutes one of the most significant footprints of the West on the rest of the world (Latour 2002: 11–12). This turns geopolitical conflicts into conflicts over space already defined, and not conflicts about the shape – or reality – of space. While the modern settlement, of course, involves much more than cartography and space, I have based this book on a narrow focus on cartography and cartographers in order to establish a historical sociology of space aiming to say something about how the historical establishment of a particular spatial reality has conditioned the political organization of space, in general, and the spatialization of the state, in particular. That is, to grant a history to space without subsuming this history to the development of the state, capitalism or related historical narratives. The cartographic transition does not in itself explain the predominance of sovereign territorial states interacting on a unified global stage, but it provided the spatial condition that made this development possible. It would not have been possible for Spain and Portugal to navigate and subject the globe the way they did without this technology to assemble the globe in Lisbon and Seville. Nor would it have been possible for territorial states to centralize and demarcate the territory the way it happened. In that respect the historical analysis has demonstrated how the so-called modern map always preceded the territory because, as has been shown, territory was very different after than before the cartographic assemblage of it. It was through cartographic means that political space was assembled as a spatial framework which was subsequently 'effectuated' through administrative centralization, infrastructural planning or worldwide trading empires.

The abstraction of space from its immediate social context, personal relations, as well as a predefined symbolic content enabled space to be organized and coordinated from centralized positions. In effect, historically, state formation and space formation entered into a mutually constitutive relationship from the 1450s to the 1650s and beyond. During this passage of time, European international relations came to be ruled by territorial states which gradually dominated over other modes of organization such as city states, the Hanseatic League, the Teutonic knights and the Church. Consequently, polities grew more similar to each other. Accordingly, identity as a political actor on a global stage converged toward a similar expression of territorial sovereignty. If one were to continue such a line of inquiry, I would expect that the increasing territorialization of politics also increased a 'demand' for social differentiation between states. This is to say that it became pertinent for the states to 'acquire' a specific 'social identity' vis-à-vis other states. Cartography both provided 'the system' with a uniform stable stage (the globe, the map of Europe, the world atlas), and it also provided each state actor with a particular spatial identity, which both provided a location within the system and also a spatial corpus. In that respect it was the map that provided the territorial state with a spatial identity to permit its constitution as a unified actor (as a political particular) on a unified global stage (as a natural universal).

The Latourian approach to space implies that space is something which is established as real through traceable social practices. Because there is no objective external reality, and likewise, no internal 'mind in a vat' that produces pure idealistic understandings of an external reality we must derive the notion of 'real' from somewhere else. In consequence, we should be looking for technological assemblages rather than spatial ontology when suggesting an epochal shift in the spatial conditions of politics. Hence, we ought to focus specifically on the practices of spatial knowledge production, and whether they are changing or not, instead of putting forward claims 'to a new spatial ontology' or a changing relationship between space and time. This should also undermine the reliance on a modern/postmodern divide regarding the spatio-temporal condition of politics. Following my argument, significant spatial changes in the territorial mode would somehow require changes to the way in which reality of space was established; either with regard to the epistemic rules or with regard to the distribution of authorship. We cannot know this from observing increasing flows of capital, deterritorialization of some security threats or increasing mobility of people unless space is made a linear function of, say, flows of capital.

In broader terms, this is an argument concerning how knowledge production of space has conditioned a specific mode of organizing politics spatially. Reading the history of cartographic practice as a history of establishing a spatial reality, it has been possible to analyse and conceptualize space as something that is both related to so-called ideational and material factors in common social science parlance. As such, the argument contributes specifically to poststructuralist and Historical Sociological understandings of the state within the International Relations (IR) discipline; adding historical agency to the former, and sensitivity to knowledge production to the latter. It is also an argument that contributes to an increasing awareness of the significance of geographical thought and practice for the development of the (post)modern world. Pickles, for example, has proposed that what he calls cartographic reason can be seen as the missing element in social theories of modernity (Pickles 2004: ix), and Peter Sloterdijk quotes Heidegger: 'The fundamental event of modernity is the conquest of the world as a picture' (Sloterdijk 2009: 29) and along similar lines than those presented by Latour, he suggests the globe as the framework for the possibility of universal thinking leading on to Heidegger's claim that the 'representation of the world with the globe is the decisive deed of the early European enlightenment' (Sloterdijk 2009: 31; for a discussion contrasting Sloterdijk's and Latour's politics, see Morin 2009). However, I also depart from these narratives in that the narrow focus on cartography and the agency behind map-making provides a more open-ended, I believe, account of the relationship between mapping, space and politics than are usually the case.

By implication of the argument that we live in a 'cartographic reality', practices of state formation and globalization are conditioned by the same spatial reality. That is, in spatial terms, there is nothing contradictory between the global and the state – on the contrary, they have come hand in hand to provide the spatial figuration of international relations. To understand the reconfiguration of political space, if this is what happens under the auspices of globalization, it is necessary to go beyond simple questions of space/not-space, territory/network or territory/global and so forth. By giving analytical autonomy to space it was possible to analyse the impact of space without either making it an absolute physical geography or a derivative of socio-economic practices. This has a number of implications for the problems concerning the literature addressing space, global change and international relations as examined in chapter two. In this analysis I emphasized three problematic distinctions that remain prevalent in the literature: first, a reliance on a subject/object divide; second, a notion of modern/postmodern historical condition of space and third, there is a persistent insistence on posing network and territory as opposites reflecting mobility and stasis.

The historical sociology of space suggested here gives a significantly different perspective on space compared to a textual reading of how meaning is written into space, or how discursive constructs write stability into a world, which can never be completely stable. This difference in methodology materialises itself as a difference in how we look for possible alternatives to the current spatial order of IR. Rather than remaining within a largely text-based analysis, which seeks openings through deconstruction, the Latourian perspective leads us to an understanding that reclaims a notion of reality without losing sight of the mundane fact that this is, of course, a construction. To determine spatial change proper, we would have to investigate the networks, and centres of calculations, within which space is produced. It is a central to this suggestion that the critical writings which associate Euclidian space with something fixed and static ignore the extent to which global space has been navigated, measured and charted according to the principles of Euclidian space in complex social processes and networks; and thus, that the apparent static nature of territorial space in fact rests on a network of spatial knowledge production. And as long as the regime of map-making remains largely the same, the spatial condition of producing territory also remains unchanged. The analysis presented here implies that it is not sufficient in itself to open a 'textual space' because the reality of space is not derived from a pure textual discourse. Instead we must unlock how space emerges as a collective between *humans* and *non-humans* to lay the course for what is possible and what is not.

When, in the literature on globalization, Jan Aart Scholte contrasts a modern three-dimensional space with a distance-less postmodern space associated with globalization he reproduces a troubled distinction. Rather than relying on a distinction between a steady modern space and a flowing postmodern space, I have shown how the formation of modern territorial states exactly took place on a stage provided by a new global cartographic reality. Indeed, the production of a global space was a pre-condition for the territorial space in its stable cartographic form. The great irony of the juxtaposition of static modern space versus postmodern accelerated spatiality is that the ability to establish static cartographic representations depended on an ability to render space mobile. Modern spatiality might be based on abstract ideals of a homogeneous, infinite and empty space, but as action does not take place in abstract space, the reality of space had to be transformed for these ideals to become a viable description of the world. And this transformation required that disparate places were made *mobile, kept stable* and *combinable*. Different sites had to be dislocated in terms of data in order to be made part of a coherent whole.

From their constructivist perspective, the effort by Ferguson and Mansbach to 'redraw our mental maps of global politics' (Ferguson and Mansbach 2004: xi) suggests that developments in culture, markets, war and technology will increasingly disembed social relations from the confines of the state and, thus, provide new constellations of political space transcending geographic territories. Yet, this focus on networked social practices vis-à-vis static territorial construction ignores the extent to which territories are also networked social assemblages, and the degree to which networks depend on and relate to territorial assemblages. Following this argument, propositions such as those advocated by Ferguson and Mansbach require more discussion of how the role of space itself changes and why it is that the market, for example, will replace the territorial logic of the state. This moves the discussion away from an assessment of the relative relevance of 'units' towards a greater concern with sociological processes of transformation. Discussions of change cannot be carried out in a language of epistemological or ontological predispositions. Instead it is necessary to trace the spatial assemblages and what role they play specifically for politics. Consequently, it is more important to investigate the constitution of 'collectives', 'units' and 'polities' rather than, for example, debating 'what the principal actors of IR are'. IR as a field of practice has always been characterized by a multitude of actors; it was private companies from Britain and the Netherlands that colonized much of Asia, and there has, if ever, only been a very short spell in history where the nation state was the only important type of actor on the global stage of politics.

The argument I have presented invariably focuses on territoriality and would appear to some as yet another state-centric inroad into the study IR and Political Geography. And in a way it is. Yet, not in the shape of Classical- or Neo-realism falling into the territorial trap as defined by Agnew (1994), but rather in order to plead for more thorough sociological analyses of the state, as well as space, before either of them is dismissed as obsolete or anachronistic. Territory in writings such as *State/Space...* (Brenner et al. 2003) appears not only as an expression of power and an existing order, though it is also that, but it is conceptualized as a more common, if not universal, feature of social relations more generally. While these writings largely remain unacknowledged in the IR literature, this area is where IR and Political Geography have a common agenda in investigating the historical trajectories and future lineages of the political organization of space in a way that avoids the straitjackets of disciplinarity and metanarratives. In *People/States/Territories*, for example, Jones argues that the agency of state personnel should occupy a more central position in our understanding of state formation and state territoriality, and he seeks to navigate between Neo-Marxist approaches which tend to neglect agency, and relational approaches that tend to neglect spatiality (Jones 2007: 13). Where Jones's exposition is right to break up the common monolithic view of the state and its link to territory, it is less attentive to the way in which different spatial realities condition these practices.

In contrast to the 'peopling of the state' as advocated by Jones, I have emphasized the agency and epistemic power of map-making, and this perspective can, I would suggest, provide some explanation as to why state territoriality remains so resilient in global politics (see also Elden 2009) despite the strong cases made for why territory should matter less. International law still refers to a cartographic spatial reality which can be seen, for example, when the United Nation's Convention on the Law Of the Sea (UNCLOS) defines the conditions under which legitimate claims can be made to maritime sovereign rights. That is, the convention defines a set of geomorphological guidelines for extending sovereign rights beyond the coast and into the ocean. §76 of the convention defines the continental shelf in terms of distance from the coast and geomorphology (Jensen 2008: 566-8), and it clearly confirms how space needs to be known in a particular way in order to become the foundation for drawing boundaries. The UNCLOS legal regime represents a fascinating attempt to move sovereign claims from the realm of (geo)politics into the scientific realm where a unified nature supposedly speaks with an unambiguous voice to politics. Hence, where we might dub this 'cartopolitics' rather than 'geopolitics' it illustrates how the international territorial order is tied up with a cartographic assemblage of space.

Turning to an issue that is closer to the disputed status of territory, it is clear that because state territory relies on a cartographic reality of space it potentially appears far removed from people's everyday experience of space as a signifier of identity, unless, of course, people have come to accept the cartographic territory as their own space. Collective identities do not necessarily correspond with the abstract spatial framework given by state territory and this tension runs through much writing on identity. A clear example is the colonial imposition of boundaries in Africa. These straight lines on the map, for the most part, corresponded poorly with the social organizations in place prior to European colonization. Yet, it was the cartographic reality of space that allowed 'colonial intervention [to freeze] African borders that had always been very mobile' (Arrous 1996: 13), and more importantly; the prevalence of cartographic reality might also explain why it is that it has proved difficult to articulate alternative spatial identities in practice within international relations generally. And for all their apparent absurdity, critique will not make such cartographic constructions vanish.

The historical global encounter between European and other cultures contained both a struggle to command 'in' space as well as a struggle to define space. And even though not all territorial spaces constructed on the basis of 'the modern global map' are accepted as appropriate foundations for political identity, geometric cartography is integrated with social organization at large. This might explain why, both at large-scale planning level but also in everyday life, maps appear as such a natural and innocent thing. Evidently, the distribution of tools such as Global Positioning Systems (GPS) and Geographic Information Systems (GIS) is unequal on a global scale, and this indicates that the cartographic reality of space is incorporated into everyday practice to a larger extent in some parts of the world compared to others. However, the key question that stands out, as a consequence of my argument, is whether there is a novel cartographic episteme emerging to replace the power of the geometric map. And here it is not enough to make references to various alternative co-existent spatial realities lived and enacted through everyday practices by smaller groups of people. Instead, to alter the epistemic power of the map, we would have to identify a cartographic episteme that establishes space in a different particular form which could replace the cartographic regime informed by geometry. To ignore this, in effect, is to ignore the epistemic power of the map and the degree to which it conditions the possibilities to be recognized as a subject in global politics. If a movement for self-determination is to be recognized, it has to be able to demonstrate the space that they claim self-determination within. Thus pointing to the map in order to claim a 'homeland' is no innocent exercise – it confirms the power of the map and the perseverance of a specific cartographic reality. This will of course constrain the ways in which the spatial identity of such a group can be articulated, and as researchers we might not find this a thing to condone, but simply to deconstruct it and call for other possibilities is to ignore the real constraints to establish a spatial identity in current politics.

The final question, then, is whether there is an epistemic change occurring today and what this would require. There is a danger that in making the argument of this book I have presented a too steadfast notion of episteme ignoring the degree to which such a knowledge system will contain openings, self-contradictions and undergo changes. Pickles, for instance, suggests that modern cartographic reason is being challenged by new cartographic forms and geometric cartography is losing its privileged position (Pickles 2004: 180–94). However, while this suggestion is based on evidence of the existence of alternative forms, inherent contradictions of cartographic reason, and a growing awareness of the power of maps, this in itself is not necessarily justifying a notion of a cartographic transition; especially not one that could justify the claims that are being put forward under the auspices of the globalization banner suggesting transformation to the political organization of space equivalent in scale and significance to that which took place between the Middle Ages and Early Modern Europe. Whether technological innovations, such as GIS, alter the traditional cartographic regime has been subject to debate. Helen Couclelis, for example, has suggested that GIS presents an opportunity to move beyond the conventional single perspective of the map in order to achieve a more appropriate representation of space (Couclelis 1999: 34–6). However, her perspective seeks to move beyond cartography and not challenge it. GIS, in her view, should not replace cartography but seek not to be captured and constrained by its rationality. And indeed developments in mapping technologies appear as a fulfilment rather than a break away from 'modern cartography'. The elevated position of the satellite represents the ideal of the divine gaze surveying the landscape that informed the transition to geometric cartography; '[a]s the eye was detached from the viewer, surveying the landscape from above, so was it presumed that the map itself was disembodied, free of human bias and prejudice, and merely mediating between a spatial reality and the viewer's perception of that reality' (Craib 2000: 20).

While this suggests that there is no immediate epistemic transition in cartography, the picture is different when we move the focus to the power of authorship. In a symptomatic fashion, the launch of the popular platform 'Google Earth' has made spatial data and images easily available to a greater public, and anyone with a laptop and Internet access can produce their own maps and locate places and buildings anywhere in the world. That this is taken seriously, and regarded with concern, is emphasized by statements that Google Earth constitutes a breach against the sovereignty of nations (Oudah 2006). This does not mean, however, that spatial censorship is a thing of the past. Following a story that insurgents in Iraq had used maps based on Google Earth imagery to locate British Army camps (Harding 2007), it has been demonstrated that Google Earth subsequently substituted contemporary satellite images of Iraq with older ones so that it should not be possible to pinpoint camps in the future.[1] Where censorship once implied the state protecting its spatial data, it is now becoming a matter of negotiations between states and large commercial companies making spatial imagery available to the public. This indicates that current changes disturb previous patterns of authorship power; however, they also suggest that traditional questions of censorship remain.

While my argument rests on a premise that there can be only one reality of space informing the international political order, one might well argue that it is exactly the existence of alternative spatial assemblages that can challenge the uniform appearance of modern international relations but the question then arises whether different spatial realities can co-exist and be part of the same spatial order. Is it possible, following Latour, to disturb the single uniform notion of nature and still have a notion of equal political communities in a unified natural world? In other words, can we maintain a formal notion of equality between polities if we move to a geopolitics which is not only about control over space but also, and more importantly, controlling the form, or reality, of space? It is central for our understanding of 'the international' to grasp the historicity of the political organization of space. The cartographic focus provides a specific contribution to established accounts on the history of the territorial state and global politics. Yet it remains for future research both to refine and expand the historical enquiries on the various spatial practices and strategies deployed by states and other actors. Not least in order to scrutinize further how, and whether, new technologies of spatial assemblage challenge the epistemic power and the power of authorship with regard to a possible new spatial reality that could challenge the current spatial order of territory, globalization and international relations.

Notes

1 Introduction

1. The notion of modern cartography is misleading and can be subject to various criticisms regarding the notion of modernity. The notions of scientific, geometrical or mathematical cartography are more precise descriptions of what is known as modern cartography, and I will use these terms interchangeably. I maintain this even though historians of cartography are likely to disagree; Brown, for example, states that '[s]cientific cartography was born in France in the reign of Louis XIV (1638–1715)' (1949: 208), and thus after the period I describe. Yet I emphasize the commencement of a move towards a cartographic theory and practice based on astronomy and mathematics even if it took centuries before these technologies were implemented and made systematic enough to be recognized as scientific in a contemporary understanding.
2. Maybe with Michel Foucault as a significant exception but he is rarely included in the canon of Historical Sociology.
3. European cartography developed from Arabic, Chinese and other knowledge traditions. The grid system, for example, informing modern cartography is known to have originated in China around the first century AD and known maps covered by such a graticule exist from the twelfth century (Turnbull 1993: 26) which is at least two centuries before the grid system came to inform European maps. For a related argument concerning European development more generally see John Hobson's *Eastern Origins of Western Civilization* (2004). Nevertheless I will abstain from detailed studies of the Chinese and other significant traditions and limit the analysis to how these practices played out in a European context.

2 The State of Territory

1. The notion of 'new mediaevalism' has inspired a whole body of literature, which hold in common, that they draw parallels between a new emerging system of international governance and, what they call, a medieval system characterized by overlapping patterns of authority (Anderson 1996; Deibert 1997; Ham 2001). Alongside the notion of a reconfiguration of authority, this literature also points to the concept of regions that, in a sense, seeks to conceptualize changes to the political organization of space. Regions represent practices that locate governance, identity and trade relations between the state and the global scale. For an overview of the literature on regions see Fawn (2009), and Paasi (2009) for a discussion of the relationship between regions and territory. The region has, especially in studies of areas within the European Union, become increasingly popular as 'a smaller geographical scale' to study than the state. Examples are Wales, Catalonia and the *Øresund* region transcending the official boundary between Denmark and Sweden. The argument

running through this literature is that regions are becoming increasingly relevant objects of study as they become more authoritative sites of production (of identity, goods, capital, etc). Related ideas are employed in the discussion of how European Space is transformed in *Organizing European Space* (Jönsson, et al. 2000) and *Making European Space* (Jensen and Richardson 2004). See Katzenstein (2005) and Buzan and Wæver (2003) for broader applications of the concept of region to the International Relations literature.

2. Several of the citations below refer to the first edition of Scholte's book. While the second edition from 2005 contains major revisions, the essence of the argument concerning space has not changed.

3. I am grateful to Jochen Kleinschmidt for articulating this critique in these terms.

3 Reclaiming a Spatial Reality

1. The term 'science wars' can be traced to C. P. Snow's classic distinction between two cultures of literature and physics that never engage in each other's spheres. With Thomas Kuhn's theory of scientific revolution, sociological factors were given a prominent place in explaining the evolution of science. These ideas were interpreted and radicalized by constructivists claiming that science was simply a socially contingent product, and therefore, did not have a privileged access to the truth (Baringer 2000: 3–9). It is this metatheoretical battle that lies at the heart of the science wars: Is science possible? Is natural science more 'scientific' than social science? And so forth. These disputes are essentially tied up with diverging notions of ontology (what exists), epistemology (how can we know about it) and anthropology (what is the human being).

2. The undermining of Euclidian geometry posed another challenge for Kant because this geometry, not only formalized the category of space as a transcendental category, but also had to describe the empirical reality. Hence, with the acceptance of a non-Euclidian geometry, it could no longer be held *a priori* that nature could be scientifically described in terms of a Euclidian space.

3. In Berger and Luckmann's classic *The Social Construction of Reality* (1991), for example, they assert how knowledge is historically contingent, and thus, specific to different societies. They operate with a dialectic between nature and the socially constructed world, where knowledge seems to operate in between. There is nature but once 'the social is established' it works back on nature. The book focuses on reality as it appears to 'the man in the street' for whom it appears immediately present and real. This reality is constructed around the here (of body-space) and now (of the moment-time). Hence, they also reproduce the distinction between a real reality, and reality as it appears to the subject located in a society that produces knowledge about reality in a historically contingent way.

4. More generally though, Emile Durkheim was early to point out how space and time are social constructions, and to make important connections between spatial construction and social organization (Durkheim and Swain 1976: 10–12). Anthony Giddens is widely credited for placing time and space at the

heart of social theory from the publication of *Central Problems in Social Theory* and onwards (Giddens 1979; Urry 1991: 161; Soja 1989: 138–56). In Giddens' (1979; 1981; 1985) writing on time-space distanciation it is emphasized how different constellations of space and time characterizes different kinds of society. However, as poignantly commented by Urry, 'Giddens' own formulations are highly frustrating, in the sense that they index some important issues but do not provide the basis for developing a really worked out position' (Urry 1991: 160). The general problem, according to Urry and Gregory, is that Giddens emphasizes time over space and that he generally fails to engage with how space, and not least different spaces, are produced (Gregory 1989; Urry 1991). In an interview with Gregory, Giddens explains (probably in a simplified fashion) how his interest in time arose from an encounter with phenomenology, and from there to an engagement with Heidegger whose notion of time-space as 'presencing' eventually made Giddens interested in space and the origins of geography (Gregory 1997: 26). This may also explain why Giddens later, in the *Consequences of Modernity*, argues that the emptying of time is a pre-condition for the emptying of space. It seems unjustified to prioritize time over space when Giddens suggests that the emptying of time is bound up with the universal measurement of time but that the emptying of space is *not* bound up with the emergence of uniform modes of measurement (Giddens 1990: 19).

5. See for example Simonsen (1996) for a good overview over developments to the mid-nineties; for prominent contributions see Keith and Pile (1993); Gregory (1994); Benko and Strohmayer (1997); Harvey (1989; 2001); Thrift (1996); Crang and Thrift (2000); Massey (2005), and Gregory (forthcoming).

6. The closest he comes to discussing the state is in 'Unscrewing the Big Leviathan...' (Callon and Latour 1981); see also Thrift (1996: 221–6), for a Latourian conceptualization of the nation state as an 'actor network'.

7. In addition to these, Heather Rae has presented an interesting alternative emphasizing what she calls 'pathological homogenization' as an integrated dimension of state formation. She poignantly shows that the issue of state formation has largely been absent from established IR theory, and the issue of identity has largely been absent from the literature on state formation. In response she combines IR theory, historical sociology and theories of cultural identity (Rae 2002).

8. 'Space' here has substituted microorganisms which are the subject of Latour's study of Pasteur's discovery of microbes.

9. I am grateful to Stuart Elden for, on a number of occasions, emphasizing this point.

4 The Cartographic Foundation of Territory

1. One of the big three post World War II historical cartographers along with Leo Bagrow (founder of the authoritative Journal Imago Mundi in 1935) and Lloyd Brown.

2. Even within the European tradition there is a linguistic heterogeneity; in many languages there are, or have been, no exclusive word for map. The word 'map'

derives from the Late Latin word 'mappa' which means cloth. The French 'carte' derives from Late Latin 'carta' and means any sort of formal document. In cultural terms, what is now considered to be maps, according to Harley's definition, existed before their creators called them maps (Harley and Woodward 1987: xvi).

3. See also Crampton (2010) for a recent overview of the literature on the specific relation between territory and cartography.

4. In addition to the *History...*project see Mundy (1996), and Lewis (1998). Furthermore, plenty of studies have shown how cartography has played a vital role in imperial and colonial projects (Edney 1997; Headley 2000; Craib 2004; Pardrón 2004).

5 The Cartographic Formation of a Global World

1. These arguments are inspired by an engaging conversation with Justin Rosenberg about his article (Rosenberg 2006) in which he draws on Wolf's work.

2. See http://avalon.law.yale.edu/15th_century/mod001.asp, last accessed 8 December 2009.

3. With the first translation of Ptolemy to Latin, geography and cosmography were used interchangeably (Headley 2000: 1120) but during the sixteenth century, Ptolemy's spatial order would be extended to a three level canon where geography became a level concerned with the earth's surface between cosmography and chorography (Cosgrove 2003).

4. It is well established that Vikings reached North America (*Vinland*), and there is also evidence which suggests that a joint Danish-Portuguese expedition might have landed in North America during the fifteenth century (Larsen 1925), although this remains speculative. On a similar note, it has been claimed that merchants from Bristol should have encountered America in the 1480s (Harley et al. 1990).

5. Some contend whether there could have been circumnavigations prior to that, and generally, the established narratives of discoveries are undergoing continuous revisions. Follow, for instance, the discussions unfolding on the Maphist discussion forum (http://www.maphist.nl/, last accessed 7 September 2010).

6. Even though Mercator's equally celebrated atlas published between 1585 and 1595 was the first to use the term, and thus give name to, the atlas.

7. It is generally difficult to maintain a meaningful distinction between what was public and private in this period but I use the distinction here to distinguish between whether the impetus of cartographic practice was part of a wider planning exercise by a centralizing state or individual entrepreneurs (for a general discussion of this, see Strandsbjerg and Kaspersen (2010)).

8. Though it should be noted that Mercator himself was opposed to the Copernican ideas and favoured the Ptolemaic system. See his correspondence with Heinrich Rantzau as discussed by Nicholas Crane (2002: 266–8).

6 The Cartographic Formation of Denmark

1. Also known by his Latin name Claudius Clavus Niger.

2. Such as *fod* (length of a foot) and *alen* (the length of the lower arm) in Denmark, these measures varied greatly in size across the regions. As any

means of calculating a standard for these were lacking, initiatives for stand-ardization came via the church, which in some cases would put templates for an *alen* on public display against which the local population could compare their own measuring tools and thus achieve a common measure within a certain area. Later on, standards for the *alen*, for example, were to be found in any city hall as ordered by royal decree in 1521 (Petersen 2002: 23–31).

3. And it seems that indirectly Lauremberg had it his way since both of the major Dutch publishing houses – Blaeu and Janssonius – began to produce new and more accurate maps of Denmark around this time. Several of these maps were dedicated to Jørgen Seefeldt, who was a friend of Lauremberg, and who could have provided the Dutch publishers with these maps (Nørlund 1943: 51).

4. The relationship between absolutism and modernity is a disputed one. By default the IR canon, which sets 1648 as a symbolic starting point of the discipline, identifies the absolutist order of post-Westphalian Europe with the modern state system. However, authors such as Teschke have disputed this and argued that one can only speak meaningfully of a modern state much later. On a similar note, Rae stresses that absolutism lay in between the medieval and the modern state (2002: 47). Among historians, writers such as Gustafsson employ the term 'conglomerate' state as a state form in between the feudal and the well-defined territorial state (2000: 25). Nevertheless, it can meaningfully be argued that many of the centralized institutions character-izing the modern state were inaugurated during the time of absolutism and, in that respect, it makes sense to relate, though not equate, the two.

7 Conclusion

1. For the analysis of the image layers on Google Earth, see 'Ogle Earth: Did Google censor Basra imagery', Sunday 14 January 2007: http://www. ogleearth.com/2007/01/did_google_cens.html. Last accessed 06/04/2010 20:24.

Bibliography

Aakjær, S. (1980). *Kong Valdemars Jordebog udg af Samfundet til Udgivelse af Gammel Nordisk Litteratur*. København, Akademisk forlag.

Agnew, J. (1994). 'The territorial trap: The geographical assumptions of international relations theory'. *Review of International Political Economy* 1(1): 53–80.

Agnew, J. A. (1998). *Geopolitics: re-visioning world politics*. London, Routledge.

Agnew, J. A. and S. Corbridge, Eds (1995). *Mastering Space: hegemony, territory and international political economy*. London, Routledge.

Akerman, J. R. (1995). 'The structuring of political territory in early printed atlases'. *Imago Mundi* 47: 138–54.

Anderson, B. (1991). *Imagined Communities: reflections on the origin and spread of nationalism*. London, Verso.

Anderson, J. (1996). 'The shifting stage of politics: new medieval and postmodern territorialities?'. *Environment and Planning A* 14: 133–53.

Angell, N. (1910). *The Great Illusion: a study of the relation of military power in nations to their economic and social advantage*. London, William Heinemann.

Aron, R. (1966). *Peace and War: a theory of international relations*. New York, Doubleday & Company, Inc.

Arrous, M. B. (1996). *Beyond Territoriality: a geography of Africa from below*. Dakar, Codesria.

Ashley, R. K. (1987). 'The geopolitics of geopolitical space: toward a critical social theory of international politics'. *Alternatives* 12: 403–34.

Ashley, R. K. and R. B. J. Walker (1990). 'Introduction: speaking the language of exile: dissident thought in international studies'. *International Studies Quarterly* 34: 259–68.

Bachelard, G. (1994). *The Poetics of Space*. Boston, Mass., Beacon Press.

Bagge, S. (1999). 'The structure of the political factions in the internal struggles of the Scandinavian countries during the high middle ages'. *Scandinavian Journal of History* 24: 299–320.

Baringer, P. S. (2000). Introduction: the 'science wars'. *After the Science Wars: science and the study of science*. K. M. Ashman and P. Barringer. London, Routledge: 1–13.

Bartelson, J. (1995). *A Genealogy of Sovereignty*. Cambridge, Cambridge University Press.

Bartelson, J. (2001). *The Critique of the State*. Cambridge, Cambridge University Press.

Bassin, M. (1987). 'Imperialism and the nation state in Friedrich Ratzel's political geography'. *Progress in Human Geography* 11: 473–95.

Baudrillard, J. (1983). *Simulations*, Semiotext[e].

Benko, G. and U. Strohmayer (1997). *Space and Social Theory: interpreting modernity and postmodernity*. Oxford, Blackwell.

Berger, P. L. and T. Luckmann (1991). *The Social Construction of Reality: a treatise in the sociology of knowledge*. London, Penguin.

Bhambra, G. K. (2007). *Rethinking Modernity: postcolonialism and the sociological imagination*. Basingstoke, Palgrave Macmillan.

Biggs, M. (1999). 'Putting the state on the map: cartography, territory, and European state formation.' *Comparative Studies in Society and History* 41(2): 374–405.

Bjørnbo, A. A. and C. S. Petersen (1904). *Fyenboen Claudius Claussøn Swart 'Claudius Clavus', Nordens ældste Kartograf*. København, Det Kongelige Danske Videnskabernes Selskab, Høst & Søn.

Black, J. (1997). *Maps and Politics*. Chicago, University of Chicago Press.

Black, J. (2000). *Maps and History: constructing images of the past*. New Haven, Yale University Press.

Blunt, A. (1968). *Artistic Theory in Italy 1450–1600*. Oxford: Oxford University Press.

BNF (1998). *Mapa Mondi: une carte du monde au XIVe siècle, l'atlas catalan* Paris, Bibliothèque Nationale de France.

BNF (1999). *Les Globes de Louis XIV: la terre et la ciel par Vincenzo Coronelli*. Paris, Bibliothèque Nationale de France.

Borges, J. L. (1998). *Collected Fictions*. London, Penguin.

Brady, T. A., H. A. Oberman and J. D. Tracy (1994). 'Introduction'. *Handbook of European History 1400–1600: late middle ages, renaissance and reformation*. T. A. Brady, H. A. Oberman and J. D. Tracy. Leiden, Brill. 1: Structures and Assertions: xiii–xxiv.

Bramsen, B. (1975). *Gamle danmarkskort en historisk oversigt med bibliografiske noter for perioden 1570–1770*. København, Rosenkilde og Bagger.

Brenner, N. (1999). 'Beyond state-centrism? Space, territoriality, and geographical scale in globalization studies'. *Theory and Society* 28: 39–78.

Brenner, N. (2004). *New State Spaces: urban governance and the rescaling of statehood*. Oxford, Oxford University Press.

Brenner, N., B. Jessop, M. Jones and G. Macleod, Eds (2003). *State/Space: a reader*. Oxford, Blackwell.

Brotton, J. (1997). 'Trading Territories: mapping the early modern world'. *Picturing history*. London, Reaktion.

Brotton, J. (1999). 'Terrestrial Globalism'. *Mappings*. D. Cosgrove. London, Reaktion: 71–89.

Brown, C., T. Nardin and N. Rengger, Eds (2002). *International Relations in Political Thought: texts from the ancient Greeks to the First World War*. Cambridge, Cambridge University Press.

Brown, L. A. (1949). *The Story of Maps*. Boston, Little, Brown and Company.

Buisseret, D. (1992). *Monarchs, Ministers and Maps: the emergence of cartography as a tool of government in early modern Europe*. Chicago, University of Chicago Press.

Buisseret, D. (2003). *The Mapmakers' Quest Depicting New Worlds in Renaissance Europe*. Oxford, Oxford University Press.

Bull, H. (1995). *The Anarchical Society: a study of order in world politics*. London, Macmillan.

Bull, H. and A. Watson (1984). *The Expansion of International Society*. Oxford, Clarendon Press.

Buzan, B. and O. Wæver (2003). *Regions and Powers: the structure of international security*. Cambridge, Cambridge University Press.

Bøggild-Andersen, C. O. (1971). *Statsomvæltningen i 1660. Kritiske studier over kilder og tradition*. Århus, Universitetsforlaget.

Callon, M. and B. Latour (1981). 'Unscrewing the big leviathan: how actors macrostructure reality how sociologists help them to do so'. *Advances in Social Theory and Methodology: toward an integration of micro and macro-sociologies*. K. K. Cetina and A. Cicourel. London, Routledge & Kegan Paul.

Cameron, A. and R. Palan (2004). *The Imagined Economies of Globalization*. London, SAGE.

Campbell, D. (1998). *National Deconstruction: violence, identity, and justice in Bosnia*. Minneapolis, Minn., University of Minneapolis Press.

Carr, E. H. (1981). *The Twenty Years' Crisis, 1919–1939: an introduction to the study of international relations : by E.H. Carr*. London, Macmillan, 1946 (1981 [printing]).

Castells, M. (2000). *The Rise of the Network Society*. Oxford, Blackwell.

Certeau, M. d. (1988). *The Practice of Everyday Life*. Berkeley, California, University of California Press.

Christianson, J. R. (2006). 'Hoffet som formidler af naturvidenskaberne under Frederik II'. *Renæssanceforum*(2): 1–16.

Clark, I. M. (1999). *Globalization and International Relations Theory*. Oxford, Oxford University Press.

Cline, H. F. (1964). 'The *Relaciones Geográphicas* of the Spanish Indies, 1577–1586'. *Hispanic American Historical Review* 13: 341–74.

Cohn, B. S. (1996). *Colonialism and its Forms of Knowledge: the British in India*. Princeton, N.J.; Chichester, Princeton University Press.

Cosgrove, D. E. (1999). *Mappings: critical views*. London, Reaktion.

Cosgrove, D. E. (2003). *Apollo's Eye: a cartographic genealogy of the earth in the western imagination*. Baltimore, Johns Hopkins University Press.

Couclelis, H. (1999). 'Space, time, geography'. *Geographical Information Systems*. P. Longley, et al. Eds. New York, John Wiley: 29–38.

Craib, R. B. (2000). 'Cartography and power in the conquest and creation of New Spain'. *Latin American Review* 35(1): 7–36.

Craib, R. B. (2004). *Cartographic Mexico: a history of state fixations and fugitive landscapes*. Durham, N.C., Duke University Press.

Crampton, J. W. (2006). 'The cartographic calculation of space: race mapping and the Balkans at the Paris Peace Conference of 1919'. *Social & Cultural Geography* 7(5): 731–52.

Crampton, J. W. (2010). 'Cartographic calculations of territory.' *Progress in Human Geography, p*republished January 28, 2010, DOI: 10.1177/0123456789123456: 1–12.

Crane, N. (2002). *Mercator: the man who mapped the planet*. London, Weidenfeld & Nicolson.

Crang, M. and N. J. Thrift (2000). 'Introduction'. *Thinking Space*. M. Crang and N. J. Thrift. London, Routledge: 1–30.

Crang, M. and N. J. Thrift, Eds (2000). *Thinking Space*. Critical geographies. London, Routledge.

Dalby, S. and G. Ó Tuathail (1998). *Rethinking Geopolitics*. London, Routledge.

Dalché, P. G. (2007). 'The reception of Ptolemy's *Geography*'. *Cartography in the European Renaissance, The History of Cartography vol. 3*. D. Woodward. Chicago University of Chicago Press: 285–364.

Dear, M. (1997). 'Postmodern bloodlines'. *Space and Social Theory: interpreting modernity and postmodernity*. G. Benko and U. Strohmayer. Oxford, Blackwell: 49–71.

Deibert, R. J. (1997). ' "Exorcismus theoriae' ": Pragmatism, metaphors and the return of the medieval in IR theory'. *European Journal of International Relations* 3(2): 167–92.

Delaney, D. (2005). *Territory: a short introduction*. Oxford, Blackwell.

Destombes, M. (1955). 'The chart of Magellan'. *Imago Mundi* 12: 65–88.

Driver, F. (1991). 'Political geography and the state formation: disputed territory'. *Progress in Human Geography* 15(3): 268–80.

DSDE (1994). *Den store danske encyklopædi*. København, Danmarks Nationalleksikon.

Due-Nielsen, C., O. Feldbæk and N. Petersen, Eds (2001). *Dansk udenrigspolitiks histori, vol. 1*. København, Danmarks Nationalleksikon.

Durkheim, E. and J. W. Swain (1976). *The Elementary Forms of the Religious Life*. London, Allen and Unwin.

Edgerton, S. Y. (1975). *The Renaissance Rediscovery of Linear Perspective*. New York, Basic Books, Inc, Publishers.

Edney, M. (2005). 'The origins and development of J. B. Harley's cartographic theories'. *Cartographica* 40(1–2): Monograph 54.

Edney, M. H. (1993). 'The patronage of science and the creation of imperial space: the British mapping of India, 1799–1843'. *Cartographica* 30(1): 61–7.

Edney, M. H. (1997). *Mapping an Empire: the geographical construction of British India, 1765–1843*. Chicago, University of Chicago Press.

Edson, E. (1997). *Mapping Time and Space: how medieval mapmakers viewed their world*. London, British Library.

Edson, E. (2001). 'Bibiographical essay: history of cartography'. *CHOICE: current reviews for academic libraries* 38 (11/12) July/August: 1899–1909.

Edson, E. (2005). 'Lost maps'. Paper presented at the International Conference of Historical Cartography in Budapest July 2005.

Edwards, C. R. (1969). 'Mapping by questionnaire: an early Spanish attempt to determine New World geographical positions'. *Imago Mundi* 23: 17–28.

Ehrensvärd, U. (2006). *The History of the Nordic Map: from myth to reality*. Helsinki, John Nurminen Foundation.

Eisenstein, E. L. (1979). *The Printing Press as an Agent of Change: communications and cultural transformations in early-modern Europe*. Cambridge, Cambridge University Press.

Elden, S. (2005a). 'Missing the point: globalization, deterritorialization and the space of the world'. *Transactions of the Institute of British Geographers* 30(1): 8–19.

Elden, S. (2005b). 'Territorial integrity and the war on terror'. *Environment and Planning A* 37(12): 2083–104.

Elden, S. (2009). *Terror and Territory: the spatial extent of sovereignty*. Minneapolis, Minn., University of Minnesota Press.

Elden, S. and J. Crampton (2007). *Space, Knowledge, Power: Foucault and geography*, edited by Jeremy Crampton & Stuart Elden. Aldershot, Ashgate.

Elliott, J. H. (2000). *Europe Divided, 1559–1598*. Oxford, Blackwell.

Ertman, T. (1997). *Birth of the Leviathan: building states and regimes in medieval and early modern Europe*. Cambridge, Cambridge University Press.

Escolar, M. (2003). 'Exploration, cartography and the modernization of state power'. *State/Space: a reader*. N. Brenner, B. Jessop, M. Jones and G. Macleod. Oxford, Blackwell: 29–52.

166   *Bibliography*

Fawn, R. (2009). '"Regions" and their study: wherefrom, what for and whereto?'. *Review of International Studies* 35 (SupplementS1): 5–34.

Fenger, O. (2000). 'Kongelev og krongods'. *Historisk Tidsskrift* 100(2): 257–84.

Ferguson, Y. H. and R. W. Mansbach (2004). *Remapping Global Politics: history's revenge and future shock*. Cambridge, Cambridge University Press.

Foucault, M. (2002). *Archaeology of Knowledge*. London, Routledge.

Foucault, M. and C. Gordon (1980). *Power/Knowledge: selected interviews and other writings, 1972–1977*. Brighton, Harvester Press.

Foucault, M., M. Senellart and A. I. Davidson (2007). *Security, Territory, Population: lectures at the College de France, 1977–78*. Basingstoke, Palgrave Macmillan.

Frye, N. and R. D. Denham (1990). *Myth and Metaphor: selected essays, 1974–1988*. Charlottesville, University Press of Virginia.

George, J. (1994). *Discourses of Global Politics: critical (re)introduction to international relations*, Basingstoke, Palgrave Macmillan.

Giddens, A. (1979). *Central Problems in Social Theory: action, structure and contradiction in social analysis*. London, Macmillan.

Giddens, A. (1981). *A Contemporary Critique of Historical Materialism, vol. 1 power, property and the state*. London, Macmillan.

Giddens, A. (1985). *The Nation-State and Violence – volume two of a contemporary critique of historical materialism*. London, Polity.

Giddens, A. (1990). *The Consequences of Modernity*. Cambridge, Polity.

Goodman, D. C. (1988). *Power and Penury: government, technology and science in Philip II's Spain*. Cambridge, Cambridge University Press.

Gottdiener, M. (1985). *The Social Production of Urban Space*. Austin, University of Texas Press.

Gottmann, J. (1973). *The Significance of Territory*. Charlottesville, University Press of Virginia.

Graham, S. (1998). 'The end of geography or the explosion of place? conceptualizing space, place and information technology'. *Progress in Human Geography* 22(2): 165–85.

Gray, C. S. (1977). *The Geopolitics of the Nuclear Era: heartlands, rimlands and the technological revolution*. New York, Crane, Russak.

Gray, C. S. (1988). *The Geopolitics of Super Power*, University Press of Kentucky.

Gregory, D. (1989). 'Presences and absences: time-space relations and structuration theory'. *Social Theory of Modern Societies: Anthony Giddens and his critics*. D. Held and J. B. Thompson. Cambridge, Cambridge University Press: 185–214.

Gregory, D. (1994). *Geographical Imaginations*. Cambridge, MA, Oxford, Blackwell.

Gregory, D. (1997). 'Space, time and politics in social theory: An interview with Anthony Giddens'. *Anthony Giddens: critical assessments*. C. G. A. Bryant and D. Jary. London, Routledge. 3: 23–35.

Gregory, D. (forthcoming). *Power Knowledge and Geography: an introduction to geographic thought and practice*. Oxford, Blackwell.

Gribbin, J. R. (2002). *Science: a history, 1543–2001*. London, Allen Lane.

Gupta, A. and J. Ferguson (1992). 'Beyond "culture": space, identity, and the politics of difference'. *Cultural Anthropology* 7(1): 6–23.

Gustafsson, H. (2000). *Gamla riken, nya stater statsbildning, politisk kultur och identiteter under Kalmarunionens upplösningsskede 1512–1541*. Stockholm, Atlantis.

Guyer, P. and A. Wood (1998). 'Introduction to the critique of pure reason'. *Critique of Pure Reason*. I. Kant. Cambridge, Cambridge University Press: 1–80.

Hale, J. R. (1971). *Renaissance Europe, 1480–1520*. London, Collins Sons & Co Ltd.

Ham, P. v. (2001). *European Integration and the Postmodern Condition: governance, democracy, identity*. London, Routledge.

Harding, T. (2007). 'Terrorists use Google maps to hit UK troops'. *The Telegraph* 13 January.

Haring, C. H. (1964). *Trade and Navigation between Spain and the Indies in the time of the Hapsburgs*. Gloucester, Mass., Peter Smith.

Harley, J. B. (1987). 'The map and the development of the history of cartography'. *The History of Cartography. Vol.1, Cartography in pre-historic ancient and medieval Europe and the Mediterranean*. J. B. Harley and D. Woodward. Chicago, University of Chicago Press.

Harley, J. B. (1988). 'Maps, knowledge, and power'. *The Iconography of Landscape: essays on the symbolic representation, design and use of past environments*. D. E. Cosgrove and S. Daniels. Cambridge, Cambridge University Press: 277–312.

Harley, J. B. (2001a). 'Deconstructing the map'. *The New Nature of Maps: essays in the history of cartography*. P. Laxton. Baltimore, MD, Johns Hopkins University Press: 149–68.

Harley, J. B. (2001b). 'Maps, knowledge, and power'. *The New Nature of Maps: essays in the history of cartography*. J. B. Harley and P. Laxton. Baltimore, MD, Johns Hopkins University Press: 51–81.

Harley, J. B. (2001c). 'Silences and secrecy: the hidden agenda of cartography in early modern Europe'. *The New Nature of Maps: essays in the history of cartography*. J. B. Harley and P. Laxton. Baltimore, MD, Johns Hopkins University Press: 83–108.

Harley, J. B., E. Hanlon and M. Warhus (1990). *Maps and the Columbian Encounter: an interpretive guide to the travelling exhibition*, American Geographical Society Collection. Milwaukee, Golda Meir Library, University of Wisconsin.

Harley, J. B. and P. Laxton (2001). *The New Nature of Maps: essays in the history of cartography*. Baltimore, MD, Johns Hopkins University Press.

Harley, J. B. and D. Woodward (1987). *The History of Cartography. Vol.1, cartography in pre-historic ancient and medieval Europe and the Mediterranean*. Chicago, University of Chicago Press.

Harste, G. (2000). 'A systemic theory of territoriality – the territorialities of European state-building'. Paper presented at the ECPR Joint Sessions of Workshops. Copenhagen 14–19 April.

Harste, G. (forthcoming). *Fra Korstog til Korstog. Krigens og Fredens Sociologi 1010–2010 i perspektiv af selvreferentielle systemer*. Århus, Aarhus Universitetsforlag.

Harvey, D. (1989). *The Condition of Postmodernity: an enquiry into the origins of cultural change*. Oxford, Basil Blackwell.

Harvey, D. (1996). 'Between space and time: reflections on the geographical imagination'. *Exploring human geography: a reader*. S. Daniels and R. Lee. London, Arnold: 443–65.

Harvey, D. O. (2001). *Spaces of Capital: towards a critical geography*. Edinburgh, Edinburgh University Press.

Headley, J. M. (2000). 'Geography and empire in the late renaissance: Botero's assignment, western universalism, and the civilizing process'. *Renaissance Quarterly* 53(4): 1119–55.

168   Bibliography

Heering, H. T. (1932). 'Knud Thott og Forhistorien til Kristian V's Matrikul'. *Tidsskrift for Opmaalings- og Matrikulsvæsen* 13(1): 1–19.
Heffernan, M. (1999). 'Historical geographies of the future: three perspectives from France, 1750–1825'. *Geography and Enlightenment*. D. N. Livingstone and C. W. J. Withers. Chicago, University of Chicago Press: 125–64.
Heiberg, S. (2006). 'Supernovaer i dansk åndsliv'. *Politiken* 3 September.
Helgerson, R. (1992). *Forms of Nationhood: the Elizabethan writing of England*. Chicago University of Chicago Press.
Helgerson, R. (1998). 'Introduction'. *Early Modern Literary Studies* 4(2): 1–14.
Henriksen, P. G. (1971). *Hærkort i Danmark og nabolande gennem tiderne*. København, Geodætisk Institut.
Henrikson, A. K. (1975). "The Map as an "Idea": The Role of Cartographic Imagery During the Second World War." *The American Cartographer* 2(1): 19–53.
Henrikson, A. K. (2002). "Distance and Foreign Policy: a Political Geography Approach." *International Political Science Review/ Revue internationale de science pol* 23(4): 437–466.
Hintze, O. (1975). *The Historical Essays of Otto Hintze*. Edited with an introduction by Felix Gilbert, with the assistance of Robert M. Berdahl, Oxford: Oxford University Press.
Hirst, P. Q., et al. (2009). *Globalization in Question: the international economy and the possibilities of governance*. Cambridge, Polity.
Hobson, J. M. (2004). *The Eastern origins of Western civilization*. Cambridge, Cambridge University Press.
Holdar, S. (1992). 'The ideal state and the power of geography the life-work of Rudolf Kjellén.' *Political Geography* 11(3): 307–23.
Howse, D. (1980). *Greenwich Time: and the discovery of the longitude*. Oxford, Oxford University Press.
Hybel, N. (2003). 'Middelalderlig godsadministration i Danmark'. *Historisk Tidsskrift* 103(1–2): 269–98.
Häkli, J. (1994). 'Territoriality and the rise of the modern state ' *Fennia* 172(1): 1–82.
Häkli, J. (1998). 'Manufacturing provinces: theorizing the encounters between governmental and popular "geographs" in Finland'. *Rethinking Geopolitics*. S. Dalby and G. O'Tuathail. London, Routledge: 131–69.
Inayatullah, N. and D. L. Blaney (2004). *International Relations and the Problem of Difference*. New York; London, Routledge.
Jacob, C. (2006). *The Sovereign Map: theoretical approaches in cartography throughout history*. Chicago, University of Chicago Press.
Jacquot, J. (1957). 'Le Théâtre du Monde.' *Revue de littérature comparée* 31: 341–72.
Jahn, B. (2000). *The Cultural Construction of International Relations: the invention of the state of nature*. Basingstoke, Palgrave.
Jammer, M. (1969). *Concepts of Space. The history of theories of space in physics*. Cambridge, Harvard University Press.
Jensen, O. B. and T. D. Richardson (2004). *Making European Space: mobility, power and territorial identity*. London, Routledge.
Jensen, Ø. (2008). 'Kontinentalsokkelens avgrensning utenfor 200 nautiske mil: Norske og russiske perspektiver i de nordlige havområder'. *Internasjonal Politikk* 66(4): 563–90.
Jespersen, K. J. V. (2004). *A History of Denmark*. Basingstole, Palgrave Macmillan.

Jespersen, M. L. (2007). 'Administration og Statsdannelse i Danmark 1400–1600'. *Den Jyske Historiker* (116): 63–87.

Jones, R. (2007). *People/States/Territories: the political geographies of British state transformation*. Oxford, Blackwell.

Jönsson, C., et al. (2000). *Organizing European Space*. London, SAGE.

Jørgensen, S. E. (2001). *Fra chaussé til motorvej det overordnede danske vejnets udvikling fra 1761*. København, Dansk Vejhistorisk Selskab i kommission hos Odense Universitetsforlag.

Kant, I. (1998). *Critique of Pure Reason*. Cambridge Cambridge University Press.

Katzenstein, P. J. (2005). *A World of Regions: Asia and Europe in the American imperium*. Ithaca, N.Y., Cornell University Press.

Katzenstein, P. J., R. Keohane and S. D. Krasner (1998). 'International organizations and the study of world politics'. *International Organization* 52(4): 645–85.

Keith, M. and S. Pile (1993). *Place and the Politics of Identity*. London, Routledge.

Kejlbo, I. R. (1980). 'Map material from King Christian the Fourth's expeditions to Greenland'. *Land- und Seekarten im Mittelalter und in der frühen Neuzeit*. C. Koeman. München, Kraus International Publications. 7: 193–212.

Keller, A. S., O. J. Lissitzyn and F. J. Mann (1967). *Creation of Rights of Sovereignty through Symbolic Acts, 1400–1800*. New York, AMS Press, Inc.

Keohane, R. (1995). 'International institutions: two approaches'. *International Theory: critical investigations*. J. Der Derian. Basingstoke, Macmillan: 279–307.

Kern, S. (1983). *The Culture of Time and Space 1880–1918*. London, Weidenfeld & Nicolson.

King, G. (1996). *Mapping Reality: an exploration of cultural cartographies*. Basingstoke, Macmillan.

Knutsen, T. L. (2007). 'En tapt generasjon? IP-fagets utvikling før første verdenskrig'. *Internasjonal Politikk* 65(3): 9–44.

Korzybski, A. C. (1948). *Science and Sanity: an introduction to non-Aristotelian systems and general semantics*. Lakeville, Conn, International Non-Aristotelian Library Pub. Co.

Kragh, H. (2005). *Dansk naturvidenskabs historie*. Århus, Aarhus Universitetsforlag.

Krarup Nielsen et al., Eds (1930). *Jordens Erobring*. København, Chr. Erichsens Forlag.

Kratochwill, F. (1986). 'Of systems, boundaries, and territoriality'. *World Politics* 39(1): 27–52.

Kula, W. (1986). *Measures and Men*. Princeton; Guildford, Princeton University Press.

Lacher, H. (2006). *Beyond Globalization: capitalism, territoriality and the international relations of modernity*. London, Routledge.

Ladewig Petersen, E. (1980). *Dansk Socialhistorie 3. Fra standssamfund til rangssamfund 1500–1700*. København, Gyldendal.

Lamb, U. S. (1974). 'The Spanish cosmographic Juntas of the sixteenth century.' *Terrae Incognitae* VI: 51–64.

Larsen, S. (1925). 'Nordamerikas Opdagelse 20 Aar før Columbus'. *Geografisk Tidsskrift* 28: 88–110.

Lash, S. and J. Urry (1994). *Economies of Signs and Space*. London, Sage.

Latour, B. (1987). *Science in Action: how to follow scientists and engineers through society*. Milton Keynes, Open University Press.

Latour, B. (1988). *The Pasteurization of France*. Cambridge, Mass., Harvard University Press.

Latour, B. (1999). *Pandora's Hope: essays on the reality of science studies*. Cambridge, Mass., Harvard University Press.

Latour, B. (2002a). 'The science wars'. *Common Knowledge* 8(1): 71–9.

Latour, B. (2002b). *War of the Worlds: what about peace?* Chicago, Prickly Paradigm Press.

Latour, B. (2005). *Reassembling the Social: an introduction to actor-network-theory*. Oxford, University Press.

Lawson, G. (2006). 'The promise of historical sociology in international relations'. *International Studies Review* 8(3): 397–423.

Laxton, P. (2001). 'Preface'. *The New Nature of Maps: essays in the history of cartography*. J. B. Harley and P. Laxton. Baltimore, ND, Johns Hopkins University Press: *ix–xv*.

Lefebvre, H. (1991). *The Production of Space*. Oxford, Basil Blackwell.

Lesger, C. (2006). *The Rise of the Amsterdam Market and Information Exchange: merchants, commercial expansion and change in the spatial economy of the Low Countries, c. 1550–1630*. Aldershot, Ashgate.

Lestringant, F. (1994). *Mapping the Renaissance World: the geographical imagination in the age of discovery*. Cambridge, Polity in association with Blackwell.

Lewis, G. M. (1998). 'Cartographic encounters: perspectives on Native American mapmaking and map use'. *The Kenneth Nebenzahl, Jr., lectures in the history of cartography*. Chicago, University of Chicago Press.

Lübcke, P. (Ed.) (1990). *Politikens Filosofi Leksikon*. København, Politiken.

Mackinder, H. J. R. (1904). 'The geographical pivot of history'. *The Geographical Journal* 23(4): 421–37.

Mackinder, H. J. R. & A. J. Pearce (1962). *Democratic Ideals and Reality – with additional papers*. New York, W. W. Norton & Company Inc.

Madsen, H. (1994). 'Der må være en kant'. *Tidsskriftet Antropologi* 30: 15–26.

Mahan, A. T. (1912). '"The great illusion".' *The North American Review* 195(676): 319–32.

Mann, M. (1984). 'The autonomous power of the state: its nature, causes and consequences'. *Archives Europeennes de Sociologie* 25: 185–213.

Mann, M. (1986). *The Sources of Social Power – volume I*. Cambridge, Cambridge University Press.

Mann, M. (1993). *The Sources of Social Power – volume II*. Cambridge, Cambridge University Press.

Marx, K. and F. Engels (1998). *The Communist Manifesto*. Oxford Oxford University Press.

Massey, D. B. (2005). *For Space*. London, SAGE.

Mead, W. R. (2007). 'Scandinavian renaissance cartography'. *Cartography in the European Renaissance, The History of Cartography vol. 3*. D. Woodward. Chicago University of Chicago Press: 1781–805.

Mignolo, W. (1995). *The Darker Side of the Renaissance: literacy, territoriality, and colonization*. Ann Arbor, University of Michigan Press.

Montesquieu, C. d. S. b. d. (1995). *De l'Esprit des lois*. Paris, Gallimard.

Morin, M.-E. (2009). 'Cohabitating in the globalised world: Peter Sloterdijk's global foams and Bruno Latour's cosmopolitics'. *Environment and Planning D: Society and Space* 27: 58–72.

Morrow, G. R. and E. Euclid (1970). *Proclus: a commentary on the first book of Euclid's Elements*. Princeton: Princeton University Press.

Mukerji, C. (1997). *Territorial Ambitions and the Gardens of Versailles*. Cambridge, Cambridge University Press.

Mukerji, C. (2006). 'Printing, cartography and conceptions of place in renaissance Europe'. *Media, Culture & Society* 28(5): 651–69.

Mundy, B. E. (1996). *The Mapping of New Spain: indigenous cartography and the maps of the Relaciones Geograficas*. Chicago, University of Chicago Press.

Neocleous, M. (2003). 'Off the map: on violence and cartography'. *European Journal of Social Theory* 6(4): 409–25.

Netterstrøm, J. B. (2007). 'Den Danske Stat og Voldsmonopolet 1440–1660'. *Den Jyske Historiker* (116): 88–119.

Nietzsche, F. W. (2003). *The Genealogy of Morals* Mineola, Dover Publications, Inc.

Nørlund, N. E. (1943). *Danmarks Kortlægning* København, Geodætisk Institut.

Nørlund, N. E. (1944a). *Islands Kortlægning: en historisk fremstilling*. København, Ejnar Munksgaard.

Nørlund, N. E. (1944b). *Færøernes Kortlægning: en historisk fremstilling*. København, Ejnar Munksgaard.

O'Brien, R. (1992). *Global Financial Integration: the end of geography*. London, [Published for] Royal Institute of International Affairs [by] Pinter.

O'Gorman, E. (1961). *The Invention of America: an inquiry into the historical nature of the New World and the meaning of its history*. Bloomington, Indiana University Press.

Ó Tuathail, G. (1996). *Critical Geopolitics*. London, Routledge.

Ó Tuathail, G. (2000). 'Borderless worlds: problematizing discourses of deterritorialization'. *Geopolitics at the End of the Twentieth Century: The Changing World Political Map*. N. Kliot and D. Newman. London, Frank Cass. 4: 139–54.

Ohmae, K. (1995). *The End of the Nation State: the rise of regional economies*. London, HarperCollins.

Ortelius, A. and R. A. Skelton (1964). *Theatrum orbis terrarum Antwerp 1570*. Amsterdam, Meridian Publishing Co.

Oudah, A.-A. (2006). 'Terrorists used Google Earth to plan attacks'. *Yemen Observer Newspaper* 31 October.

Padrón, R. (2004). *The Spacious Word Cartography, Literature, and Empire in Early Modern Spain*. Chicago, University of Chicago Press.

Pagden, A. (1993). *European Encounters with the New World: from renaissance to romanticism*, Yale University Press.

Painter, J. (2006). 'Territory-network'. Paper presented at Association of American Geographers Annual Meeting. Chicago 7–11 March.

Painter, J. (2008). 'Cartographic anxiety and the search for regionality'. *Environment and Planning A* 40(2): 342–61.

Palan, R. (2003). *The Offshore World: sovereign markets, virtual places, and nomad millionaires*. Ithaca, N.Y., Cornell University Press.

Pardrón, R. (2004). *The Spacious Word – cartography, literature, and empire in early modern Spain*. Chicago, University of Chicago Press.

Parker, G. (1992). 'Maps and ministers: the Spanish Habsburgs'. *Monarchs, Ministers and Maps: the emergence of cartography as a tool of government in early modern Europe*. D. Buisseret. Chicago, University of Chicago Press: 124–52.

Parry, J. H. (1990). *The Spanish Seaborne Empire*. Los Angeles, University of California Press.

Parry, J. H. (2000). *The Age of Reconnaissance 1450–1650*. London, Weidenfeld & Nicolson.

Petersen, K. (2002). *Mål og vægt i Danmark*. Lyngby, Polyteknisk Forlag.

Pickles, J. (2004). *A History of Spaces: cartographic reason, mapping, and the geo-coded world*. London, Routledge.

Pottage, A. (1994). 'The measure of Land ' *Modern Land Review* 57: 361–384.

Price, D. J. (1955). 'Medieval land surveying and topographical maps'. *The Geographical Journal* 121(1): 1–10.

Ptolemy, n. c. et al. (2000). *Ptolemy's Geography: an annotated translation of the theoretical chapters*. Princeton, Princeton University Press.

Paasi, A. (2009). 'The resurgence of the "region" and "regional identity": theoretical perspectives and empirical observations on regional dynamics in Europe'. *Review of International Studies* 35(SupplementS1): 121–46.

Rae, H. (2002). *State Identities and the Homogenisation of Peoples*. Cambridge, Cambridge University Press.

Rasmussen, C. P. (2007). 'Delene og Helheden: statsdannelse i det dansk-ledede monarki ca. 1380–1700'. *Den Jyske Historiker* (116): 34–62.

Ratzel, F. (1896). 'The territorial growth of states'. *Scottish Geographical Magazine* 12(7): 351–61.

Revel, J. (1991). 'Knowledge of the territory'. *Science in Context* 4: 133–61.

Ringmar, E. (1996). 'On the ontological status of the state'. *European Journal of International Relations* 2(4): 439–66.

Rosenau, J. N. (2004). 'Many globalizations, one international relations'. *Globalizations* 1(1): 7–14.

Rosenberg, J. (2000). *The Follies of Globalisation Theory: polemical essays*. London, Verso.

Rosenberg, J. (2006). 'Why is there no international historical sociology?' *European Journal of International Relations* 12(3): 307–40.

Ruggie, J. G. (1993). 'Territoriality and beyond – problematizing modernity in international-relations'. *International Organization* 47(1): 139–74.

Ruggie, J. G. (2004). 'Reconstituting the global public domain – issues, actors, and practices'. *European Journal of International Relations* 10(4): 499–531.

Russell, J. B. (1991). *Inventing the Flat Earth: Columbus and modern historians*. New York; London, Praeger.

Sack, R. D. (1980). *Conceptions of Space in Social Thought: a geographic perspective*. London, Macmillan.

Sack, R. D. (1986). *Human Territoriality: its theory and history*. Cambridge, Cambridge University Press.

Sahlins, P. (1989). *Boundaries: the making of France and Spain in the Pyrenees*. Berkeley, Oxford, University of California Press.

Said, E. W. (1995). *Orientalism*. London, Penguin.

Sandman, A. (2007). 'Spanish nautical cartography in the renaissance'. *Cartography in the European Renaissance*. D. Woodward. Chicago, University of Chicago Press. 3: 1095–142.

Sassen, S. (2000). 'Territory and territoriality in the global economy'. *International Sociology* 15(2): 372–93.

Sassen, S. (2006). *Territory, Authority, Rights*. Princeton, Princeton University Press.

Scafi, A. (1999). 'Mapping Eden: cartographies of the Earthly Paradise'. *Mappings*. D. Cosgrove. London Reaktion.

Scholte, J. A. (1996). 'Beyond the buzzword: towards a critical theory of globalization'. *Globalization: theory and practice*. E. Kofman and G. Youngs. London, Pinter: 43–57.

Scholte, J. A. (2000). *Globalization: a critical introduction*. London, Macmillan.

Scholte, J. A. (2005). *Globalization: a critical introduction*. Basingstoke, Palgrave Macmillan.

Scocozza, B. (2003). *Ved afgrundens rand*. København, Gyldendal & Politiken.

Scott, J. C. (1998). *Seeing Like a State: how certain schemes to improve the human condition have failed*. New Haven, Conn., Yale University Press.

Serchuk, C. (2006). 'Picturing France in the fifteenth century: The map in BNF MS Fr. 4991'. *Imago Mundi* 58(2): 133–49.

Shapiro, M. J. (1994). 'Moral geographies and the ethics of post-sovereignty' *Public Culture* 6: 479–502.

Shields, R. (1999). *Lefebvre, Love and Struggle: spatial dialectics*. London, Routledge.

Short, J. R. (2004). *Making Space Revisioning the World, 1475–1600*. Syracuse, N.Y., Syracuse University Press.

Simonsen, K. (1996). 'What kind of space in what kind of social theory?' *Progress in Human Geography* 20(4): 494–512.

Skelton, R. A. (1965). *Decorative Printed Maps of the 15th to 18th Centuries. A revised edition of Old Decorative Maps and Charts by A.L. Humphreys. With ... a new text by R.A. Skelton*. London, Spring Books.

Skelton, R. A., H. Hexham and H. Hondius (1968). *Mercator-Hondius-Janssonius: Atlas or A geographicke description of the world*. Amsterdam, Theatrum Orbis Terrarum.

Sloterdijk, P. (2009). 'Geometry in the colossal: the project of metaphysical globalization'. *Environment and Planning D: Society and Space* 27: 29–40.

Smail, D. L. (2000). *Imaginary Cartographies: possession and identity in late medieval Marseille*. Ithaca, Cornell University Press.

Smith, M. L. (2005). 'Networks, territories, and the cartography of ancient states'. *Annals of the Association of American Geographers* 95(4): 832–49.

Smith, S. (2004). 'Singing our world into existence: international relations theory and September 11'. *International Studies Quarterly* 48: 499–515.

Soja, E. W. (1971). 'The political organization of space'. *Commission on College Geography Resource Paper* Washington DC, Association of American Geographers. 8.

Soja, E. W. (1989). *Postmodern Geographies: the reassertion of space in critical social theory*. London, Verso.

Sparke, M. (2005). *In the Space of Theory: postfoundational geographies of the nation-state*. Minneapolis, Minn., University of Minnesota Press.

Stanley, H. M. (1885). 'Inaugural address – delivered before the Scottish Geographical Society at Edinburgh, 3rd December 1884'. *The Scottish Geographical Magazine* 1: 1–17.

Storey, D. (2001). *Territory: the claiming of space*. Harlow, Pearson Education.

Strandsbjerg, J. and L. B. Kaspersen (2010). 'De territoriale forudsætninger for offentligt-privat samarbejde'. *Offentligt og privat – historiske og aktuelle betragtninger*. L. B. Kaspersen, J. Lund and O. H. Petersen, Eds. Copenhagen, DJØF Forlag.

Szakolczai, A. R. D. (1998). *Max Weber and Michel Foucault: parallel life-works*. London, Routledge.

Taylor, P. J. (1994). 'The state as container: territoriality in the modern world-system.' *Progress in Human Geography* 18(2): 151–62.

Taylor, P. J. (2000). 'Geopolitics, political geography and social science'. *Geopolitical Traditions: a century of geopolitical thought*. K. Dodds and D. Atkinson. London, Routledge: 375–9.

Teschke, B. (2003). *The Myth of 1648: class, geopolitics, and the making of modern international relations*. London, Verso.

Teschke, B. (2006a). 'Debating "the myth of 1648": state formation, the interstate system and the emergence of capitalism in Europe – a rejoinder'. *International Politics* 43: 531–73.

Teschke, B. (2006b). 'Geopolitics'. *Historical Materialism* Leiden, Koninklijke Brill. 14: 327–35.

Thrift, N. (2000). 'Its the little things'. *Geopolitical Traditions: a century of geopolitical thought*. K. Dodds and D. Atkinson. London, Routledge: 380–7.

Thrift, N. (2003). 'Space: the fundamental stuff of geography'. *Key concepts in geography*. S. L. e. a. Holloway. London, SAGE Publications: 95–108.

Thrift, N. J. (1996). *Spatial Formations*. London, SAGE.

Thrower, N. J. W. (1999). *Maps & Civilization – cartography in culture and society*. Chicago, University of Chicago Press.

Tilly, C. (1992). *Coercion, Capital and European States, A.D.990–1990*. Oxford, Blackwell.

Tunander, O. (2001). 'Swedish-German geopolitics for a new century – Rudolf Kjellén's "The state as a living organism"'. *Review of International Studies* 27(3): 451–63.

Turnbull, D. (1993). *Maps are Territories: science is an atlas*. Chicago, University of Chicago Press.

Turnbull, D. (1996). 'Cartography and science in early modern Europe: mapping the construction of knowledge spaces'. *Imago Mundi* 48: 5–24.

Turnbull, D. (2000). *Masons, Tricksters and Cartographers: comparative studies in the sociology of scientific and indigenous knowledge*. Newark, Harwood Academic.

Ulsig, E. and A. K. Sørensen (1981). 'Studier i Kong Valdemars Jordebog – Plovtalsliste og Møntskat'. *Historisk Tidsskrift* 81: 1–25.

Urry, J. (1991). 'Time and space in Giddens' social theory'. *Giddens' Theory of Structuration: a critical appreciation*. C. G. A. Bryant and D. Jary. London, Routledge: 160–75.

Urry, J. (2000). *Sociology Beyond Societies: mobilities for the twenty-first century*. London, Routledge.

Urry, J. (2007). *Mobilities*. Cambridge, Polity.

Venge, M. (1981). 'Clementsfejden og Caspar Paludan-Müller.' *Historie – Jyske Samlinger* Ny Række XIV(1): 1–36.

Walker, R. B. J. (1993). *Inside/Outside: international relations as political theory*. Cambridge, Cambridge University Press.

Waltz, K. N. (1979). *Theory of International Politics*. New York, McGraw-Hill Inc.
Weber, M. (1994). 'The profession and vocation of politics'. *Weber: political writings*. P. Lassman and R. Speirs. Cambridge; New York, Cambridge University Press: 309–69.
Wendt, A. (1999). *Social Theory of International Politics*. Cambridge, Cambridge University Press.
Wendt, A. (2006). '*Social Theory* as Cartesian science: an auto-critique from a quantum perspective'. *Constructivism and International Relations: Alexander Wendt and his critics*. S. Guzzini and A. Leander. London, Routledge: 181–219.
Whitfield, P. (1994). *The Image of the World: 20 centuries of world maps*. London, British Library.
Winichakul, T. (1996). 'Maps and the formation of the geo-body of Siam'. *Asian Forms of the Nation*. S. Tønnesson and H. Antlöv. Surrey, Curzon Press.
Wolf, E. R. (1990). *Europe and the People without History*. Berkeley; London, University of California Press.
Wolff, T., Ed. (1979). *Det matematisk-naturvidenskabelige Fakultet – 2.del*. Københavns Universitet 1479–1979. København, G-E-C Gads Forlag.
Wood, D. (1992). *The Power of Maps*. New York, The Guilford Press.
Woodward, D. (1991). 'Maps and the rationalization of geographic space'. *Circa 1492: art in the age of exploration*. J. A. Levenson. New Haven, Yale University Press: 83–7.
Woodward, D. (2007a). 'Cartography and the renaissance: continuity and change'. *Cartography in the European Renaissance*. D. Woodward. Chicago, University of Chicago Press. 3: 3–24.
Woodward, D., Ed. (2007b). *Cartography in the European Renaissance*. The History of Cartography. Chicago, University of Chicago Press.
Wæver, O. (1996). 'The rise and fall of the inter-paradigm debate'. *International Theory: positivism & beyond*. S. Smith, et al. Eds. Cambridge, Cambridge Press: 149–85.
Østergård, U. (2002). 'The state of Denmark – territory and nation'. *Comparare* 2: 1–14.

Index

Page numbers in **bold** refer to figures.

182 *Index*